# THE MOON

## An observing guide for backyard telescopes

By
MICHAEL T. KITT

From the publishers of
**ASTRONOMY**
Magazine

Kalmbach Books, Waukesha, Wisconsin

# *Foreword*

I can't figure out what happened. It was 20 years ago that human beings last set foot on the Moon. But until now no lunar guide book for amateur astronomers incorporated the scientific findings from Apollo missions and others. Those missions revolutionized all of our understanding of the Moon—where it came from, what made its craters, when its vast lava plains oozed forth. But little of these results has crept into the observing guides for amateur astronomers. It's sort of like a book on cosmology neglecting to tell you about the Big Bang.

To put it bluntly, up until now backyard Moon observers have been shortchanged. Mike Kitt's book is the first lunar observing guide that explains the Moon's real geologic history. No longer is the Moon a blank sheet on which any old idea can be projected. The real Moon is a fascinating rocky desert that now has an accurate history. And it's the only extraterrestrial body you can explore in such detail.

Mike Kitt takes you on a month-long tour of the lunar landscape and tells you how it really got that way. In this delightful book I have re-encountered features that are old friends and discovered other ones that I had never noticed. If you're a newcomer to astronomy—welcome! Here's the lunar story. If you're an old-timer, get ready for a new look at some familiar sights.

As the United States contemplates a return to the Moon, every telescope owner can feel a growing excitement in learning more about this neighbor world right on our celestial doorstep. This book provides a perfect way to begin exploring the wonders of the lunar globe. And once you know how it all happened, you'll never again look at that starkly beautiful landscape in the same way.

Robert Burnham
Editor, ASTRONOMY

# THE MOON
## An observing guide for backyard telescopes

**ASTRONOMY LIBRARY NO. 1**

Kitt, Michael, 1941-
    The moon : an observing guide for backyard telescopes / Michael Kitt.
       p.  cm.
    ISBN 0-913135-09-7 (pbk.)
      1. Moon--Observer's manuals.    I. Title.
  QB581.K52   1991
  523.3--dc20                     91-62557

To Matthew and David
with hope for the future;
and to Margie, who lives
in Eternity's sunrise

# 1. How to Observe the Moon

The Moon is delightful in that you can obtain pleasing views of it with almost any optical aid, from an inexpensive pair of binoculars to a large Newtonian telescope or high-quality refractor. But to appreciate fully the many fine details on the Moon, you should take a small number of steps to set yourself up to best advantage. Using an appropriately outfitted telescope is certainly a key factor in carrying out lunar observations. Beyond that, successful observers also know how to judge whether atmospheric conditions are acceptable for high-resolution viewing, when the best times for observing the Moon are, and what techniques will provide for a pleasurable and productive session at the eyepiece.

## Observing Equipment

An excellent telescope for lunar observation would be a 5- or 6-inch refractor or an 8- to 10-inch reflector with a focal ratio of at least f/7. Smaller instruments also work well if they possess good optics and are, in the case of refractors, well color-corrected. A well-made 4-inch f/15 refractor or a 6-inch f/8 reflector would leave little to be desired. Huge apertures, common with the Dobsonian-mounted Newtonians of today, can be a liability when observing the Moon. You gain a little perhaps in resolution, but they collect so much light that they overwhelm you with glare that is awkward to reduce by filters. However, telescopes with apertures exceeding 14 inches can be used if you outfit them with an off-axis mask to reduce the effective aperture to 6 to 10 inches. Naturally, they should have good optics that are well collimated and free of aberrations, just as for other aspects of observational astronomy.

Equally important is that the instrument be mounted on a stable equatorial mount, preferably equipped with a clock drive. Lunar and planetary observation is mostly high-power work; plus you'll be spending long periods at the eyepiece waiting to catch moments of superb seeing and searching out fine details that may escape notice at first glance. A mounting and drive that makes it easy to keep the Moon centered in the field under high magnification is absolutely imperative. Unfortunately, this means that Dobsonian and other altazimuth mounted telescopes are next to useless for prolonged lunar observation, no matter how fine their optics. While these telescopes will give marvelous glimpses of the rugged lunar terrain, using them to take a close look at even an easy target such as the Alpine Valley will simply lead to frustration as the Moon rapidly glides through the field of view. I suspect this operational difficulty has discouraged many people from really discovering what the Moon has to offer. Fortunately, some excellent equatorial platform designs for Dobsonian mounts have been developed. These offer a tracking system accurate enough to convert these telescopes for lunar and planetary use.

Eyepieces are also important when it comes to lunar observing. The Moon provides excellent viewing opportunities at low, moderate, and high magnification. Low power observation provides a global view of the lunar surface, showing the interrelationship of seas and highland areas and displaying the vast lunar ray systems. Generally, low-power viewing involves magnifications of about 60x. A good orthoscopic, Plössl, or Kellner design eyepiece in the range of 20mm to 25mm will serve very well for this purpose. (The exact focal length you need will depend on your telescope's focal length.) The wide-field eyepiece designs used by deep-sky observers don't offer any particular advantage when used for lunar observing, nor is there any need for 2-inch barrel eyepieces; the standard inch and a quarter models work fine.

Moderate power observing, usually carried out at magnifications of about 150x, provides detailed regional views of the Moon. At this power, you can explore most of the major formations with the expectation of spotting some of their smaller details. This is also a fine magnification range to use for identifying specific targets to be explored at higher powers. A good orthoscopic, wide-field, or Plössl eyepiece in the 12mm to 15mm range will provide excellent viewing.

Lunar observing is carried out at high powers whenever conditions permit. The global and regional views discussed above are an essential and esthetic prelude to any observing session, but the critical details that can provide a sense of discovery are best revealed at powers of 300x and more. While orthoscopic design eyepieces have traditionally been used and provide acceptable views, the Nagler 4.8mm and 7mm eyepieces provide a quantum leap in image quality. Any serious observer of the Moon should obtain at least one of these for high-resolution lunar observations.

A high-quality achromatic Barlow is a fine supplement for providing extreme magnification on superb nights. Contrary to the guidelines often cited for general observing, lunar and planetary observers can successfully use magnifications of 600x and higher on those infrequent occasions when the atmosphere is unusually steady. For example, an 8-inch f/7 reflector used with a 4.8mm Nagler and a 2x Barlow will provide a magnification of nearly 600x. The special techniques required for observing at these high magnifications will be discussed below.

Though it reflects only 7 percent of the light falling on it — about the same as fresh asphalt — the Moon is an extremely bright object in the nightime sky. At low and moderate magnifications you will feel uncomfortable — and overlook detail — unless something is done to reduce the Moon's glare. A polarizing filter set (about $30) usually proves the most efficient answer to the problem. By rotating one of the two filters with respect to each

**A very slender "new" Moon hangs low in the western sky after sunset. Its welcome reappearance like this each month marks the beginning of another viewing period for lunar observers. D.L. Coleman photo.**

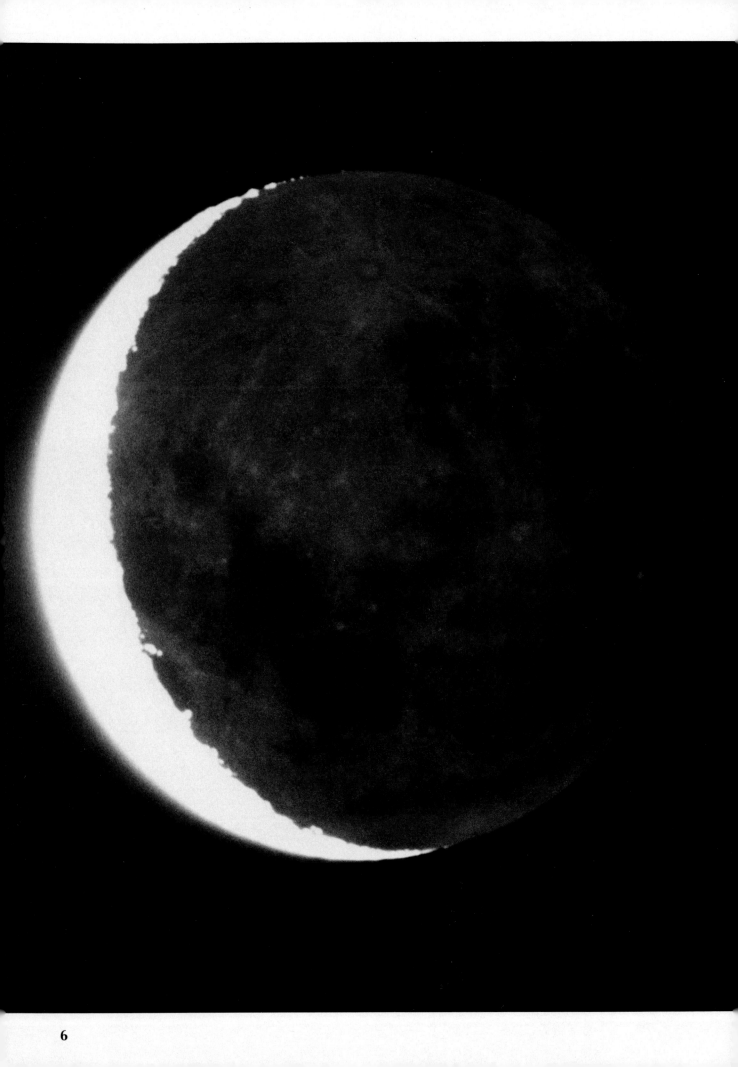

other, you can adjust the exact amount of light reduction you need for comfortable viewing.

The Moon will retain a natural appearance until the filters are nearly fully crossed. At this point (which you would use only for extreme light reduction) you'll see the Moon pick up a blue cast. A less expensive alternative is to view through a neutral density filter (roughly $15 each). These are typically sold as "Moon filters" and models are available that screw into the bottom of the eyepiece. At moderate magnifications, many observers find a deep yellow filter can be effective and has the property of improving contrast for certain lunar formations.

## Observing Conditions

For most kinds of observing, we think of sky transparency as being the most important criterion in determining whether or not to break out the telescope. This is not true when it comes to observing the Moon. It's plenty bright enough! The key criterion in lunar observation is good *seeing* or sky steadiness.

In fact, nights of superb transparency more often than not turn out to be marginal for viewing the Moon. Dark nights usually come on the heels of a cold front's passage and these seldom produce settled seeing. In many locations, the nights of steadiest seeing often feature a slight amount of haze. For planet viewing, light haze can begin to interfere with observations. But the Moon is so bright and contrasty that thin hazes are rarely detrimental. Some of the best lunar observing nights I can remember occurred when it was difficult to see stars much below 4th magnitude. Of course, you shouldn't automatically write off crisply clear nights. When you get a dark night *and* settled seeing, then you've got the ultimate in viewing conditions — and a night you'll long remember.

Also keep in mind that good seeing can change during the night, even rapidly. An evening of marginal steadiness, good for low-power views only, can suddenly and dramatically improve, apparently without reason and with no accompanying change in weather or transparency. Therefore, don't give up on the evidence of a few minutes' of unsatisfactory viewing. Once you have set up, leave the scope in place and recheck conditions periodically to see if improvements have occurred. You may be in for a delightful surprise.

Good seeing also has a local dimension. Every observing site, even within a given area, can have certain periods during the night and times of the year when observing conditions are more favorable. The best times during the night, whether early evening or near midnight, can also vary on a seasonal basis. Experience will show you which places and times are best for you. But even neophytes can help improve their seeing by choosing an observing site wisely. Avoid viewing over houses or pavement; trees, grass, or shrubs are better because they don't absorb as much heat during the day. A driveway or rooftop will be radiating heat far into the night — heat that will keep in motion image-destroying air currents.

Also be realistic: on some nights, the atmosphere just won't settle down enough to permit long drawn-out views of the Moon at high magnification. When this happens,

don't battle nature. Just back off on the magnification to the point where the seeing no longer calls attention to itself. And keep in mind, too, that even a night with slight turbulence will be punctuated by short intervals of rock-steady seeing. If you see this happening, make yourself comfortable at the eyepiece, monitor the lunar surface at high power, and concentrate on detail only when periods of steadiness occur.

Better seeing also comes when the Moon is higher above the horizon. As a general rule, it's best to wait until the Moon is at least 45° high for the most critical high-power viewing. And the higher the Moon, the better. When the Moon is low in the sky, we look through a significantly larger slice of Earth's atmosphere.

## When to Observe the Moon

Naturally, viewing a 3- or 4-day old Moon doesn't give you much choice when to observe — at this phase the Moon never climbs high above the horizon (except during full daylight). But at other times during the lunar month, you can plan observing sessions for a time when the Moon is riding as high in the darkened sky as possible. Generally speaking, until the Moon reaches First Quarter, you are compelled to view it in the evening. As the Moon approaches Full, you can view later and later in the evening. By the time it reaches Last Quarter, you'll have to view after midnight, and observing an old waning crescent Moon means catching it soon before sunrise.

Casual looking is something you can enjoy anytime, no matter where the Moon is. But if you want to see the Moon when it lies highest in the sky, there are optimum times of year for observing specific phases. For observers in the north temperate zone, the best time to view the Moon is when it lies in the zodiacal constellations of Pisces, Aries, Taurus, Gemini, Cancer, and Leo. For each phase of the Moon, this placement occurs at a specific time of year. Thus a 4-day old Moon will be highest above the horizon at twilight in the month of May. The First Quarter Moon will be found in Taurus/Gemini and at highest elevation in the month of March. The Full Moon rides high overhead at midnight in late December, and a Last Quarter Moon is best seen in September. (Conversely, if you go out to observe a First Quarter Moon in September, you'll find it in the Scorpius/Sagittarius region, where it will never attain much altitude above the horizon.) Viewers in the southern hemisphere should add six months to each date given.

But don't take these rules as ironclad — you'll find acceptable viewing conditions for each lunar phase for several months around these optimum dates. And of course, there's nothing stopping you from viewing the Moon whenever it's visible. Another aspect of the lunar orbit has a small effect on observing conditions, but it's important for those formations located near the limb of the Moon. This has to do with libration, a phenomenon which causes the Moon to appear to rock back and forth slowly as seen from Earth. Because of libration, we can actually see 59% of the Moon's surface at one time or another. And as a result of libration, formations lying close to the lunar limb will periodically swing more toward the center of the disk, improving the view we have of them.

There are actually three independent kinds of libration. The first, libration in longitude, results in the Moon's swaying back and forth in the east-west axis, allowing us to alternately see as much as 8° beyond the nominal east and west limbs. This form of libration is caused by a slight

**Sunlight reflecting from Earth's oceans and clouds paints the shadowed portion of the moon with a ghostly radiance called earthlight. (Unless noted south is at top to match the view in a telescope.) Cecil H. Johnson, Jr. photo.**

mismatch between the Moon's constant rotation speed around its axis (its "day") and its varying velocity in orbit around the Earth (its "month"). If the Moon travelled in a perfectly circular orbit around the Earth, we would always see precisely the same lunar face. But the Moon actually travels in an elliptical orbit and therefore its orbital speed varies. The Moon moves faster than it rotates when it is closest to Earth and more slowly when it is farther away. This means that at one point in its orbit, the Moon's rotation is lagging a little behind the orbital velocity and two weeks later rotation gets a little ahead. These variations cause the Moon to appear to slowly rock back and forth in the east-west axis as seen from Earth.

The second type of libration, libration in latitude, causes the Moon to appear to rock north and south. This occurs because the Moon's axis of rotation tilts 6.5° to the plane of its orbit around Earth. From our viewpoint, the north pole of the Moon will occasionally tip toward us (and the south pole tip away) and two weeks later the reverse happens.

The third libration takes place each day between moonrise and moonset, when our planet's rotation carries the observer a distance of 4,000 miles — one Earth-radius — from one side of the Earth-Moon line to the other.

Librations sometimes work at cross-purposes and at other times they augment each other. Librations can be predicted using complicated mathematical models. But few observers bother — it's much simpler to judge the state of the libration by merely scanning the Moon at low power and observing the orientation of three easy-to-spot formations. These are Mare Crisium in the east, the crater Plato in the north, and the crater Grimaldi in the west. All three are dark in color and easy to locate.

Routine observation of Mare Crisium with binoculars over the course of a few months will show you the effects of libration in longitude. One night, let's say you notice that Mare Crisium nearly touches the eastern limb of the Moon. Then over the next two weeks, it will gradually move more onto the disk so that a broad swath of bright highland area lies between the dark mare and the eastern edge of the Moon. Then it will slowly return to its former location, perhaps going just as far or perhaps not, depending on whether librations are working in sync or out of it.

At the same time, noting the changing distance of the crater Plato from the apparent lunar north pole will give you a measure of libration in latitude. For example, suppose you want to observe the crater Pythagoras, which is magnificent but located not too far from the lunar limb in the northwest quadrant of the Moon. When the librations are unfavorable, seeing any details in Pythagoras can be highly difficult because the crater will appear close to the limb. Sunrise on Pythagoras occurs three days before Full Moon. On that night, let's say you take a quick look at Plato and Mare Crisium. (Grimaldi, the third benchmark, will still be hidden by the lunar night.) If Plato lies well away from the northern edge of the Moon, you can count on a decent view of Pythagoras that evening because libration in latitude is favoring the Moon's northern hemisphere. And if at the same time, Mare Crisium appears close to the eastern limb, conditions for observing Pythagoras will be close to ideal because it will indicate that li-

**A rugged landscape becomes evident by First Quarter (left), roughly a week after New Moon. This, the Moon's eastern hemisphere, contains five large "seas" which actually are vast plains of frozen lava. Above: A slow wobble in the Moon's rotation, which is called libration, ensures that we do not always see exactly the same face from month to month. Lick Observatory photos.**

bration in longitude has moved the northwest limb closer toward the center of the lunar disk.

When you view features well away from the limb, libration has no significant effect on their visibility. But for objects in the outer third of the Moon's radius, libration can have a profound effect on what you see and how well you see it. Formations located very close to the limb can swing completely over the edge and out of sight! A quick scan at the beginning of an observing session will be enough to tell you the state of affairs as far as libration is concerned for that evening. Then it is simply a matter of concentrating on the portion of the limb that is favorably displayed.

## Observing Technique

One of the first things anyone notices about the Moon when they glance in a telescope is how stark the scenery looks. The lack of a lunar atmosphere gives shadows sharply defined edges. Consequently, on the lunar surface things are either brightly lit or totally in darkness. There is no twilight on the Moon. This type of lighting presents an enormous advantage to the lunar observer. The sharp detail and extreme contrast inherent in the lunar surface let you resolve features far smaller than would be possible for a lunar astronomer peering at the Earth. On the other hand, the Moon is essentially a monochrome body — it has virtually no color variations, save for a few subtle hints of brownish and greenish tints. Thus the Full Moon, lit from overhead and devoid of shadow, reveals only broad details involving albedo (brightness) features like the dark maria and the bright lunar rays.

Observing the Moon means exploring those areas where on a given night, the terrain is thrown into high relief by low-angled sunlight. In general, this means that the majority of observing will be concentrated in a strip with

in about 15° of the lunar terminator, the line between night and day. The terminator sweeps across the disk twice in each lunar month. (Did you know that at the lunar equator the terminator moves at about 8 kilometers per hour, or 5 miles an hour — you could keep pace with it on foot!). Since the terminator advances across the Moon about 12° each night, successive evenings each bring out a wholly different set of formations to be seen at their best.

For the vast majority of lunar formations, most of the detail is lost only a day after local sunrise. A few of the freshest craters with strong features, such as Tycho and Copernicus, continue to have enough shadowing to show details for several days before sunset or after sunrise. But those formations that are smooth and have little relief — the lunar domes and wrinkle ridges — must be observed when the terminator is as close to them as possible.

Albedo features (consisting of bright rays, dark spots on the floors of craters, and the maria themselves) are generally best seen when the Sun is at least halfway to the zenith — or, put another way, about four or more days after the sunrise terminator has passed across them. By this time, virtually all shadows will have disappeared in the area. At Full Moon, the entire disk is dominated by albedo features, and identifying most craters will be beyond the abilities of even the most experienced observer.

**Last Quarter Moon is dominated by two large lava flows, Oceanus Procellarum and Mare Imbrium, which together cover much of the western lunar hemisphere (right). Above: The Moon's distance from Earth varies over the course of a month. The change in apparent size shows clearly in photos taken at perigee and apogee using the same equipment. Photos by Russ Sampson (above) and Lick Observatory (right).**

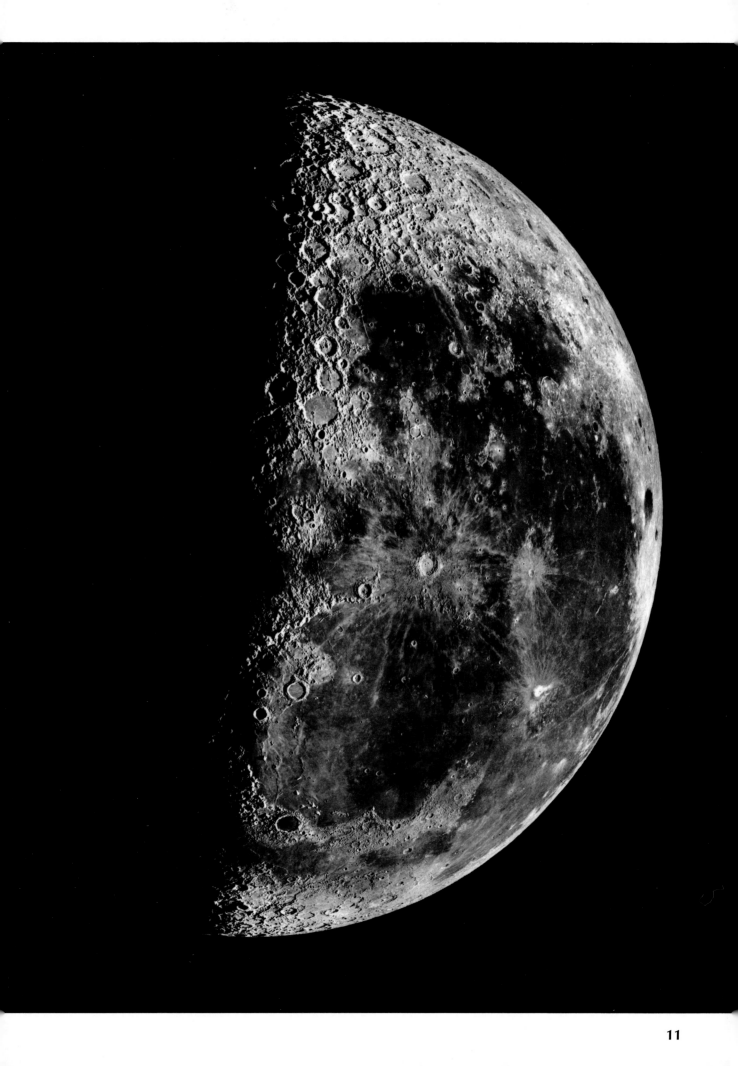

## Typical Lunar Observing Session

Now let's lay out a useful plan for observing the Moon. If you previously decided to look at a specific formation, perhaps you marked your calendar, knowing that it would be visible so many days after New Moon. The length of your observing session can be as long or as short as you desire. Half an hour might suffice if you want to check one or two formations — and an entire evening can slip by if sky conditions are superlative.

I first look up at one or two bright stars. If they are twinkling madly, it's likely a waste of time to even set up, although it is always wise to check again in a half-hour or so to see if the turbulence has diminished. Immediately after setting up the telescope or opening the observatory, I use a low-power eyepiece and check the exact position of the terminator by identifying any major formation along its length. I also check the state of librations to see if any portion of the sunlit limb is well placed for observation. Then while the telescope optics are cooling to equilibrium with the air, I plan my observing session — determining which formations should be well placed and reading up about them.

When the telescope has cooled, my practice is to begin observing using moderate magnification of about 150x. First, I test for atmospheric steadiness by noting the sharpness of the image and looking right at the edge of the sunlit lunar limb. If turbulence is present, the limb will appear to shimmer. How can you tell the difference between atmospheric turbulence and optics which have not properly cooled down? In general, the tipoff is that if it's the optics, the shimmering will follow a cyclical pattern, whereas atmospheric effects tend to be more random. At moderate powers, the Moon should appear steady and crisp. However, slight amounts of turbulence are tolerable if your observing plan is to become familiar with a region of the Moon and to practice finding and identifying formations.

Next I scan the length of the terminator, starting at one end near the limb of the Moon and moving to the other. I look for the specific formations that are particularly well lit at that moment, taking time to appreciate the details. Usually, at least half a dozen formations show excellent detail under the lighting conditions at any given time. I find I don't need filters with an 8-inch telescope at 160x, provided that the field is centered on the terminator. However, if I move in toward the sunlit portion of the Moon where the local Sun is higher, the amount of light brightens too much for comfortable viewing.

Having surveyed the terminator and gotten a good look at the target objects for the evening, I center one of them and begin attacking it with progressively higher powers. Moving up to about 250x, I begin looking for increased details within the formation. If the seeing is holding up well, I next increase the magnification to about 400x. At this magnification, the lunar surface never looks as crisp as it does at 150x. Yet fine details completely invisible at the lower power will become apparent if the seeing is good enough. On the finest nights, out comes my Barlow, and I make observations at 500x to 800x for as long as conditions permit.

At this point, I should caution you about over-magnification. Given the right eyepieces, you can pump up the power of a telescope to the point where a small crater will fill the field of view and look monstrous. If viewing conditions are excellent, details will still be visible even at enormous magnification. But most times this approach results in what experienced observers call "empty magnification," which means that you are increasing the apparent size of the object without resolving any additional detail. Indeed, you are actually *decreasing* the amount of detail that the eye can perceive. The key in moving up to a higher magnification is to examine a particular detail for a few minutes and determine whether or not the view is improved. If not, back off to a lower power.

However, before you reduce magnification, investigate one other possibility. On occasion atmospheric turbulence can vary, with moments of complete steadiness that come and go. Therefore, having decided to use a lower power eyepiece, I nonetheless continue observing at high power for a minute or two. Sometimes the view steadies for a few seconds, providing a momentary razor-sharp view. If enough of these moments of excellence make it worthwhile, I will often sit gazing through the eyepiece, waiting for the seeing to steady up, then concentrate on observing detail in those few precious moments. Sometimes this is the only practical way to observe at extreme magnifications, since local conditions can cause long periods of bad seeing regularly.

Having determined the maximum useful magnification, I now proceed to move through the list of formations for the evening, always alert for improving or deteriorating seeing. If the session lasts for a couple of hours, I return to moderate power and scan the Moon to see if the advancing terminator has revealed new formations. After completing my high-power survey of the Moon, I always look at the lunar panorama at low power, providing a full-disk view. With the aid of glare-reducing filters, I examine the lunar ray systems and the dark areas far from the terminator. I always scan the circumference of the limb for dark mare areas brought into view by favorable librations.

If the Moon is a crescent, I also make it a point to examine the earthlit side of the disk, using an unfiltered low-power eyepiece and being sure to keep all of the brightly lit lunar surface outside the field of view. Like soft ghosts, some of the brightest and darkest areas will be identifiable, including the various lunar seas. Don't expect to see much detail, but be prepared for a most esthetically pleasing view.

The secret of successful and enjoyable lunar observation is preparation and selectivity. Those who observe the Moon simply because it's up in the sky will probably become bored quickly. Select the best nights for your observations, and then pick the best formations to view —those whose shadows are placed to show off all their details. A few minutes reviewing the formations you will be looking at, using the information contained in section 3, will amply repay the time you spend on it. If anyone doubts the importance of this preparatory review, I invite them to perform the following experiment. Go out on a clear night and examine any prominent formation visible on the Moon. Take careful note of the various details you see. Now open this book and read about that formation. I predict that more often than not, you will discover that you have overlooked some details — details that will immediately jump into view when you return to the eyepiece.

The main thing, though, is to have some pleasureable moments using your telescope. The Moon is a whole world at your disposal. It has scenic wonders, stark beauty, and mystifying formations. You can observe the Moon to try to pierce the veil of these mysteries, or you can just turn off the clock drive and pretend to be circling in orbit above the forbidding lunar landscape. All you need to do is say, "I think I'll set up my scope tonight!"

# 2. A Guide to Lunar History

Previous generations have looked at the Moon and wondered at the origin of the landscape before them. They often misunderstood what they saw. Lunar observing consists in large part of viewing the shadows cast by formations on its surface. Our normal day-to-day experiences here on Earth tell us that interpreting shadows is risky business. So it's no surprise that as recently as the 1950s, highly experienced lunar observers could not agree on the true nature of the Moon's most common feature, its craters.

Humans have now walked on the Moon and orbited above it. As a result, many of its mysteries have been solved. For the first time we can observe the lunar landscape with an educated eye. Features formerly passed over have taken on new significance in the light of what has been learned about lunar geology. And we are no longer fooled by intriguing-looking but insignificant formations that were once the centers of controversy.

The panorama presented to us through the eyepiece is enormously complex. But to an astonishing degree you can actually see many of the Moon's most important features with a small telescope from your own backyard. The Moon has not yielded all of its secrets and many more significant discoveries are yet to be made. But today's amateur astronomer who learns something of the Moon's history and geology will be in a position to view the Moon in ways no one ever could before. All you need is clear vision and a different point of view.

## The Evolution of the Moon

The modern picture of our Moon's birth and its subsequent evolution towards the starkly beautiful object we see today is the product of the golden age of lunar exploration. This began in July 1964 with the Ranger 7 mission and ended in August 1976 with the return of lunar samples by Luna 24. During this twelve-year period Apollo astronauts visited the Moon six times. These expeditions brought back to Earth extensive collections of lunar rocks, gravel, and dust, which were supplemented by three successful Soviet automated sample-return missions. Five U.S. Surveyor missions and two Soviet Lunokhod missions obtained samples for further analysis of the lunar surface. These were supplemented by many photographic projects, most notably the five Lunar Orbiter and eight Apollo missions.

In the time since these data and samples were gathered, scientists have forged a coherent explanation of the Moon's geological evolution. This new understanding starts with the Moon's violent birth some 4.6 billion years ago and divides lunar history into five distinct eras. The **Pre-Nectarian Era** ran from the formation of the Moon until 3.92 billion years ago, when the Nectaris basin impact occurred. It encompasses the period when the Moon accreted, melted throughout, and chemically differentiated. The next period, classified as the **Nectarian Era**, lasted from 3.92 billion years ago to 3.85, and is a period

during which the last remaining large planetesimals in this region of the solar system slammed into the Moon and blasted out the great impact basins.

The impact that created one of these basins, now occupied by Mare Imbrium, lends its name to the **Imbrian Era**. It began with the Imbrium impact 3.85 billion years ago and lasted 700 million years; it is a period characterized by extensive volcanism. There then followed a long period of quiescence known as the **Eratosthenian Era**, named for a medium-sized crater which typifies the cratering which occurred during this era. Beginning 3.15 billion years ago and ending about 1.2 billion years ago, the Eratosthenian Era encompasses half the Moon's history. Finally we come to the **Copernican Era**, named for the great crater whose name it bears. It spans a little more than the last billion years of lunar history. This era, continuing down to modern times, is marked by minimal activity and is best known for the formation of relatively fresh craters with splashy ray systems.

**The Pre-Nectarian Era.** The Moon was born 4.6 billion years ago in an episode of violence typical of the early solar system. As scientists reconstruct it, a Mars-size planetesimal gave the early Earth a glancing blow. This spewed a mixture of Earth material and planetesimal pieces into orbit around our planet. In the blink of a geological eye, these fragments coalesced into the proto-Moon, which continued to grow as more ejecta from this impact and the large amounts of debris still left over from the formation of the solar system fell onto the Moon's surface.

Scientists think the Moon formed quickly from the accretion of primordial solar system material, the entire process taking perhaps only 60 to 80 million years. The last stages of this titanic drama were so rapid and violent that the energy released by the impacts melted the Moon. The melting may also have been aided by heat from the decay of short-lived radioisotopes such as aluminum-26 and by strong tidal interactions with the Earth. When the Moon melted, it differentiated. The heavy materials sank toward the core and lighter minerals floated upward to form the crust. Samples of the lunar crust are enriched with radioactive isotopes of potassium, uranium, and thorium. If the whole Moon had the same concentration of elements, the radioactive decay alone would have been enough to melt the Moon completely.

The rock samples returned from the Moon's highlands suggest that a global ocean of melted rock, or magma, was created during this episode. The lunar crust is largely composed of the light-colored mineral plagioclase feldspar (also called anorthosite); the crust averages 60 kilometers in thickness. The only way such an enormous amount of anorthosite could have formed would have been by the existence of a global magma ocean extending 400 kilometers or more into the Moon. As this ocean cooled, low-density plagioclase crystals floated upwards and concentrated in a layer which became the lunar crust. Denser materials, such as olivine and pyroxene, sank to form what later became the source zone for the dark lavas we now see in the lunar "seas."

The total absence of water and low levels of sodium and potassium in lunar rocks suggests that the magma ocean must have been extremely hot, causing these volatile substances to boil off into space. This hot magma ocean cooled until a skin of plagioclase formed. Gradually the zone of freezing rock migrated deeper and by the end of the Pre-Nectarian Era, 3.92 billion years ago, a substantial crust was in place. Over the ensuing ages this molten zone generated mare lavas which came from deeper and deeper within the Moon, until the molten zone was so deep that lavas could no longer reach the lunar surface. Today, the molten zone lies at a depth of about 1,100 kilometers, only 638 kilometers from the Moon's center.

**The Nectarian Era.** One of the biggest scientific surprises to come from the Apollo program was the knowledge that the heavily cratered lunar highlands were not relics of the accretion process, but actually formed more than 500 million years after the Moon was largely complete. The Nectarian Era, encompassing the period between the Nectaris basin impact (3.92 billion years ago) and the Imbrium basin impact (3.85 billion years ago), represents the time during which the highlands formed. This era is characterized by periods of violent meteoritic bombardments which ended in an event now referred to as the terminal cataclysm or the late heavy bombardment.

When lunar highland rocks were first chemically dated, planetary scientists noted that many dates clustered around an age of 3.9 billion years. Scientists had expected that these rocks would date back to the time of their formation, that is, to when the magma ocean solidified. This would have given the rocks an age of at least 4.2 billion years. Clearly something had reset the radiometric clocks in these rocks.

The most plausible explanation is that massive shocks occurred when most of the circular basins visible on the Moon were blasted out by huge impacts. The fact that highland rock samples obtained from widely separate locales have nearly identical ages points to the notion that the impacts occurred within a short period of time. The enormous volumes of ejecta from these basins must have blanketed most of the Moon.

It is clear that the intense bombardment of the terminal cataclysm abruptly ceased about 3.9 billion years ago at the end of the Nectarian Era, because rocks formed shortly after this time are found well-preserved. By that time, some 43 basin-making impacts distributed evenly across the lunar globe and countless smaller ones had deeply pulverized the Moon's crust. A layer of rubble at least a mile or more deep (called the mega-regolith) had been formed, and fragments of different rocks were cemented together to form a composite kind of rock called a breccia. These breccias were in turn broken apart, tossed great distances, and re-cemented together with new fragments, creating a legacy of highly complex rocks with histories that are difficult to decipher.

**The Imbrian Era.** An outstanding event forms a benchmark in the Moon's early history. About 3.85 billion years ago an asteroid about 100 kilometers in diameter struck the Moon with devastating results. It excavated the Imbrium basin, some 1,500 kilometers across. A thick blanket of ejected material, now called the Fra Mauro Formation, was thrown over much of the near side of the Moon. The Fra Mauro ejecta moved as if it were partly

fluid and pooled in low-lying crater floors and in the intercrater plains. Elsewhere, huge gouts of ejected rock scoured the vicinity, truncating the Haemus mountains to the east. Farther away, flying boulders gouged radial valleys still visible to this day. In a few short hours the face of the Moon was drastically altered.

These were just the superficial, visible effects of the impact. The lunar crust was also deeply shattered, right down to the molten zone, which at that time was perhaps 150 kilometers below the surface. This shattering of the crust initiated the great age of lunar volcanism known as the Imbrian Era. No doubt there was some volcanic activity prior to the Imbrium impact. But after it, large quantities of lava welled to the surface through a multiplicity of cracks and chasms. The molten rock slowly filled most of the nearside basins left over from the Nectarian Era. As the lava froze in place, it formed the lunar maria we see today. Curiously, there are no significant maria on the lunar farside — perhaps because the focus of the shattering force of the Imbrium impact was concentrated on the nearside of the Moon.

The lava plains of the maria were built up gradually by successive eruptions, which spread for hundreds of kilometers. As the Apollo 15 astronauts saw when they gazed into Hadley Rille, the lava seas of the Moon are composed of multiple thin layers, each only a few meters thick. Samples of the lava show that when molten they had the consistency of thin motor oil; in contrast, on Earth lava is commonly much thicker. Its runny consistency permitted the lunar lava to flow great distances before congealing. Lunar lava was much too fluid to build volcanic shields such as those found on Earth, Mars, and Venus. The fluid lavas also erased much of the evidence of volcanic activity — the vent craters and the lava tubes — by ponding at the very end of the eruptions. However, as your telescope will show you, many rilles and craters clearly volcanic in origin can still be seen today.

It would be a mistake to envision the Imbrian Era as a time of relentless volcanic eruptions. In fact, since the lunar seas contain only a relatively small volume of lava, it is clear that over the 700-million-year era lengthy periods of quiescence must have occurred. As time passed, the zone of molten rock sank progressively deeper towards the core of the Moon. Lava that came from a depth of 150 kilometers at the beginning of the era was ultimately coming from 400 kilometers below the surface by its close. The chemical composition differences of the lava are apparent from variations in shade and tint visible through a telescope.

When basaltic flows filled the lunar basins, the lava lay heavily on the anorthositic crust it covered. Unable to support the great weight of the basalts, the crust fractured and began subsiding in the centers of the basins. The fractures provided new conduits for additional lava flows. Evidence for tectonic fracturing is preserved in the form of concentric rille systems found on the perimeters of many of the lunar seas, especially on the eastern shore of Mare Humorum near the crater Hippalus. Another type of tectonic adjustment may have created the mare wrinkle ridges, gently uplifted folds on the surfaces of the seas.

**Streaked with light and dark, the Full Moon is dazzling. There's no better time to explore its ray systems — splashes of powdered and melted rock flung from the Moon's newest and freshest impact craters. Lick Observatory photo.**

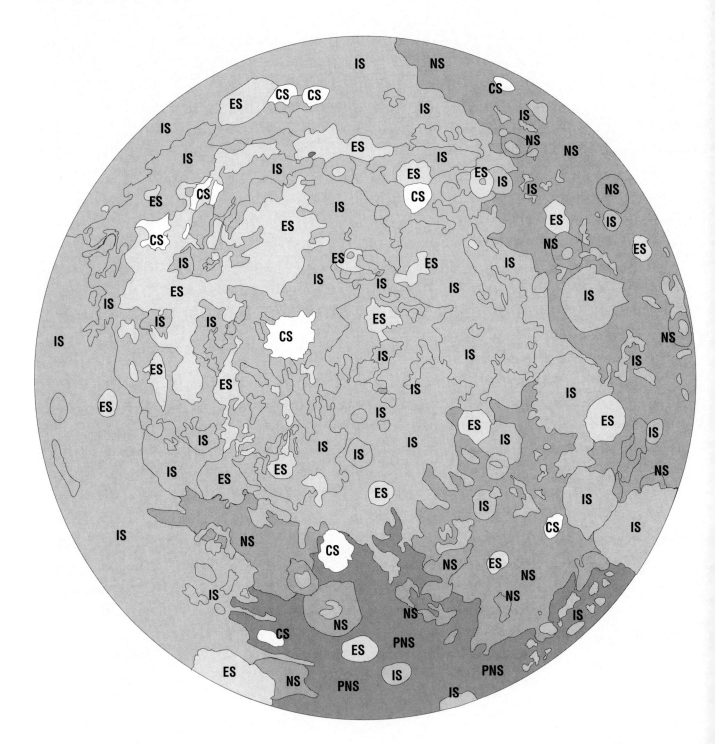

Copernican System (CS)

Eratosthenian System (ES)

Imbrian System (IS)

Nectarian System (NS)

Pre-Nectarian System (PNS)

**Impact occurs 3.9 billion years ago**

© Don Davis

The cooling of the Moon continued until the end of the Imbrian Era. At that point, about 3.15 billion years ago, the crust had thickened enough to effectively block significant further migration of molten lava to the surface.

**The Eratosthenian Era.** If you could view the Moon as it looked at the beginning of the Eratosthenian Era roughly 3 billion years ago, its general appearance would be similar to today. Except for changes in the bright rays around the freshest craters — and the absence of a couple dozen moderate-sized craters such as Tycho and Copernicus — the Moon's maria and highlands have not changed all that much since widespread volcanism ceased. In this sense it's correct to say that the Moon is essentially a dead world. And yet even rare geological events on inactive solar system bodies can have significant cumulative effects when they are given billions of years to work.

The Eratosthenian is the longest period in lunar history, beginning 3.15 billion years ago and ending 1.2 billion years ago. Ample evidence suggests that low level volcanism must have continued from Imbrian time well into the Eratosthenian Era. Crater counts in selected areas, for example near the crater Lambert in Mare Imbrium and with in the Marius dome field in Oceanus Procellarum, show this clearly. We can also point to a number of small- to moderate-sized craters which date to this era.

But the most profound changes that occurred throughout the two-billion-year duration of the Eratosthenian Era were caused by the inexorable pelting of the Moon by tiny meteoroids. Of course, small impacts had been common in all periods of lunar history. But with the cessation of both large, frequent impacts and widespread lava flooding, these celestial mosquito bites finally came into their own. Systematically they pulverized and abraded the Moon's outermost skin, throwing rock fragments long distances in the low gravity.

The end result of this process is evident in the many surface photographs obtained during the lunar missions. Almost everything the astronauts saw, photographed, picked up, and walked upon had a rounded and subdued appearance. If you remember pre-Apollo paintings of the Moon's surface, you can't help but be struck by the absence of jagged peaks or sharp-edged rocks on the real Moon. Slowly but very persistently impact "gardening" over countless ages has resulted in erosion of all but the most recent of lunar formations.

**Impact creates shock wave, heats surface**

**The Copernican Era.** The last 1.2 billion years of the Moon's evolution is grouped into a separate era simply because the features from this period are well preserved and maintain a fresh appearance. From a purely geological point of view, however, there is little to distinguish the Eratosthenian Era from the Copernican.

This latest era, which continues down to the present, is typified by the craters Copernicus (810 million years old) and Tycho (109 million years). Both craters are well preserved and are the centers of large, magnificent ray systems which dominate the appearance of the Full Moon. In time, the continuing impacts of tiny meteoroids will degrade the appearance of these craters.

But well before that happens, their ray systems will fade. Sputtering by solar wind ions and bombardment by small cosmic particles will slowly erase these surface features. Millions of years are required to wear off a layer only a fraction of a millimeter thick. But the rounded upper surfaces of lunar rocks ejected from impact craters only 10 to 20 million years old show that the process, while slow, is undoubtedly efficient.

**Today's Moon.** The structure of the Moon today is rea-sonably well known from seismic studies. The lunar crust is about 60 kilometers deep on the nearside; on the farside it may be as thick as 100 kilometers. Meteoritic bombardment has turned the upper layers of the crust into a rubble called regolith. The regolith is perhaps two kilometers deep in the highland regions and generally 5 to 20 meters deep on the maria. The uppermost surface material is the finely divided lunar soil.

Beneath the lunar crust is a thick mantle, now solid and rigid to depths of 600 to 800 kilometers. Moonquakes are routinely recorded at these depths, indicating that deformation and tectonic adjustments still occur. The zone from which the mare lavas derived, extending from 150 to 400 kilometers deep, has long since solidified. In 1972, the impact of a large meteorite provided seismic data demonstrating there is a zone of molten rocks about 1,200 kilometers down. Because few seismic events are strong enough to send shock waves through the core of the Moon, less is known about it. It is at least partially molten and probably accounts for the inner 200 to 700 kilometers of the Moon's 1728-kilometer radius. If the core is composed of iron sulfide, its size would be closer to the

**Material cools into shallow basin**

© Don Davis

700 km figure. However, if the core is nickel-iron like Earth's, 200 kilometers is probably the maximum size.

The Moon's present seismic activity is low — typical moonquakes reach a Richter scale value of only 2. (Someone once calculated that a medium-sized Fourth of July fireworks display gives off more energy than does the whole Moon in a year!) These lunar disturbances occur most often when the Moon is at perigee and apogee and are thus related to Earth's tidal forces. Moonquakes occur deep in the interior, some 600 to 800 kilometers below the surface. This is in contrast to terrestrial earthquakes, which mostly occur within the upper 200 kilometers. About 40 moonquake centers have been identified and most of these lie on the nearside.

Given the thickness of the solidified crust and mantle, it is doubtful that lunar volcanos have erupted for some time. Yet there is ample evidence that sporadic venting of gases regularly occurs, perhaps at the time of the tidally generated moonquakes. Radon gas has been detected near the crater Aristarchus and around the edges of many of the circular mare. And numerous telescopic observations of red glows and temporary obscurations, so-called lunar transient phenomena or LTPs, are, if genuine, most likely related to the release of small quantities of gas, which may raise dust and fluoresce before dissipating into the vacuum surrounding the Moon.

In over 350 years of telescopic observation of the Moon, no one has ever seen a new crater or other formation being formed — although the history of lunar study is littered with disproven claims to that effect. Of course, this is perfectly consistent with our experience on Earth, for any projectile large enough to create an observable crater on the Moon would also be large enough to penetrate the Earth's atmosphere and leave a clear footprint. And since there are few fresh impact craters on Earth, we can conclude that in modern times such events are fairly infrequent.

## Formations on the Moon

Without knowing anything at all about the origins of the lunar features, you can happily gaze at them with a telescope for a lifetime. But if you know even a little about the formations, you'll find it adds immeasureably to the pleasure you'll have while observing them. The lunar ob-

**Basin floods with molten basaltic rocks, 2.7 billion years ago**

© Don Davis

server of today, benefiting from the knowledge gained by space exploration, "sees" far more than his or her counterpart of the past. Details become more apparent when you realize they are not random, but instead have a genetic relationship to other features in the field of view.

We have already touched upon some types of lunar formations in reviewing the Moon's evolution. There are the impact features: lunar craters ranging in size from gargantuan basins to the smallest bowl-shaped craters we can see. Then mountain ranges and scarps, which in reality are the remnant walls of ancient basins and craters. Valleys scoured out by impact ejecta, and the lunar ray systems associated with the freshest craters. Finally, there are the volcanic and tectonic features: the lunar maria with their associated wrinkle ridges, rilles, and domes. Let's now look at these types of features in more detail.

**Lunar Craters.** The most basic lesson learned from the space program is that impact cratering has been a pervasive and fundamental force throughout the solar system. The Moon owes the greatest part of its topography to the effects of impact cratering at all scales. Meteoroids typically impact at speeds of many kilometers per second.

When this nearly irresistable force meets the almost immoveable Moon, solid rock can explode with enormous energy. The tremendous kinetic energy of a high-speed collision is converted on impact into thermal, acoustic, and mechanical energy capable of inflicting great damage on the impact site. The great kinetic energy resulting from solid objects moving at cosmic speeds enables relatively small objects to produce gigantic craters.

The first event to occur in crater formation is compression. When a meteoroid hits the Moon, it penetrates the surface and is quickly engulfed. Violent shock waves propagate through the meteorite and the target. But seismic waves can travel only at about 12,000 miles per hour, whereas the meteorite is moving three to four times faster. Thus it is impossible for the rock to dissipate the impact energy via seismic waves. The result is that in the initial microseconds after impact, both the meteorite and the target are intensely compressed. Some parts of both objects are melted and vaporized.

At the same time this crushing is happening, fluidized rock particles are jetting away from the impact site at high speed. The material can travel at speeds exceeding

**Basaltic rock cools into Mare Imbrium**

© Don Davis

90,000 miles per hour. Much of it flies out into space and is lost forever, but some may fall back to the surface to make a delicate filamentary ray system that can stretch for thousands of kilometers. The compression stage, which generally lasts less than 1 second, ends when the shock wave reflects off the back of the already deformed meteorite.

The second stage of crater formation is the excavation of the crater itself. Shock and decompression waves carry away the energy of the impact and set material in motion. This explosive event excavates from the impact site a volume of material much larger than the meteorite. This creates a crater that may be 20 to 50 times the diameter of the projectile. This excavated material, or ejecta, is flung in a blanket-like layer outwards by about one crater diameter. The ejecta blanket then becomes ropy and discontinuous and gradually thins out. Isolated groups of secondary craters form when blocks of rock thrown high out of the crater come crashing down. The entire process lasts from 10 seconds to a few minutes, depending on the size and force of the impact.

The final stage of crater formation consists of post-cra-

tering modifications. These include adjustments to the floor of the crater when it rebounds from the force of the impact. This rebound may create a single central peak or groups of them. Landslides and slumps can occur on the crater's inner walls, creating complex terraces at various heights above the floor and depositing landslide material on the floor itself. You can often see that the intense fracturing of the bedrock beneath the crater tapped underground reservoirs of magma, which then seeped up to flood the crater's floor.

Impact craters are generally circular or polygonal but may take on a variety of other shapes depending on the composition and speed of the projectile and the structure of the impact area. But the most significant factor determining a crater's attributes is the projectile's size. Very large craters can look so different from small ones that it was once thought different processes formed them. This led to a now-obsolete crater classification system that invoked names such as "walled plains" and "ring mountains" for the largest examples. You'll see such evocative terms in the older writings about the Moon, but they have now been abandoned because they tend to obscure what actually happened.

**Slow erosion and cratering have produced today's lunar surface**

Nonetheless, impact craters fall into at least three major groups, based on attributes related to size (and hence to the force of the impact that created them). The largest craters, those created by the impacts of asteroid-sized bodies, are called multi-ring basins. Very large impacts shatter significant amounts of terrain around the impact zone. The effect is much like the spreading rings created by dropping a pebble into water. It is difficult to recognize the multi-ringed basins on the nearside of the Moon because subsequent volcanism has covered much of the evidence. Photographs of the lunar farside, where little volcanism occurred, reveal that all the larger impacts resulted in formations with two, three, or more concentric rings. The prime example is Mare Orientale, which straddles the Moon's western limb. Once farside photos alerted them, planetary scientists reexamined the Moon's nearside and found ample evidence that all the lunar basins were once multi-ringed.

Below a diameter of about 300 kilometers, impact craters do not exhibit multiple rings, but instead the flow of solid material results in a central mountainous region. For many older craters in this size range, the central peaks have been destroyed by volcanism and subsequent impacts or buried under the ejecta from nearby impacts that pooled over them.

Finally, craters with diameters of less than about 20 kilometers exhibit none of the structural features we have mentioned and are merely smooth concave holes in the ground, still surrounded by ejecta from the impact. These are called bowl craters, and this simple form of crater is maintained down through the smallest size visible in a telescope to the smallest microcraters on the lunar surface.

Any crater can be modified after impact, which has the effect of creating other classes of craters. The most common are the so-called flooded craters. These are ordinary craters that have been invaded by lava flows, which flooded their floors and destroyed their central peaks and other surface detail. Examples include the craters Plato and Archimedes. Occasionally the flooding is so extensive that the crater interior fills up with lava, as has occurred with Wargentin and Posidonius. More frequently, molten lava outside the crater will breach the wall on one side. This results in beautiful lunar bays such as Sinus

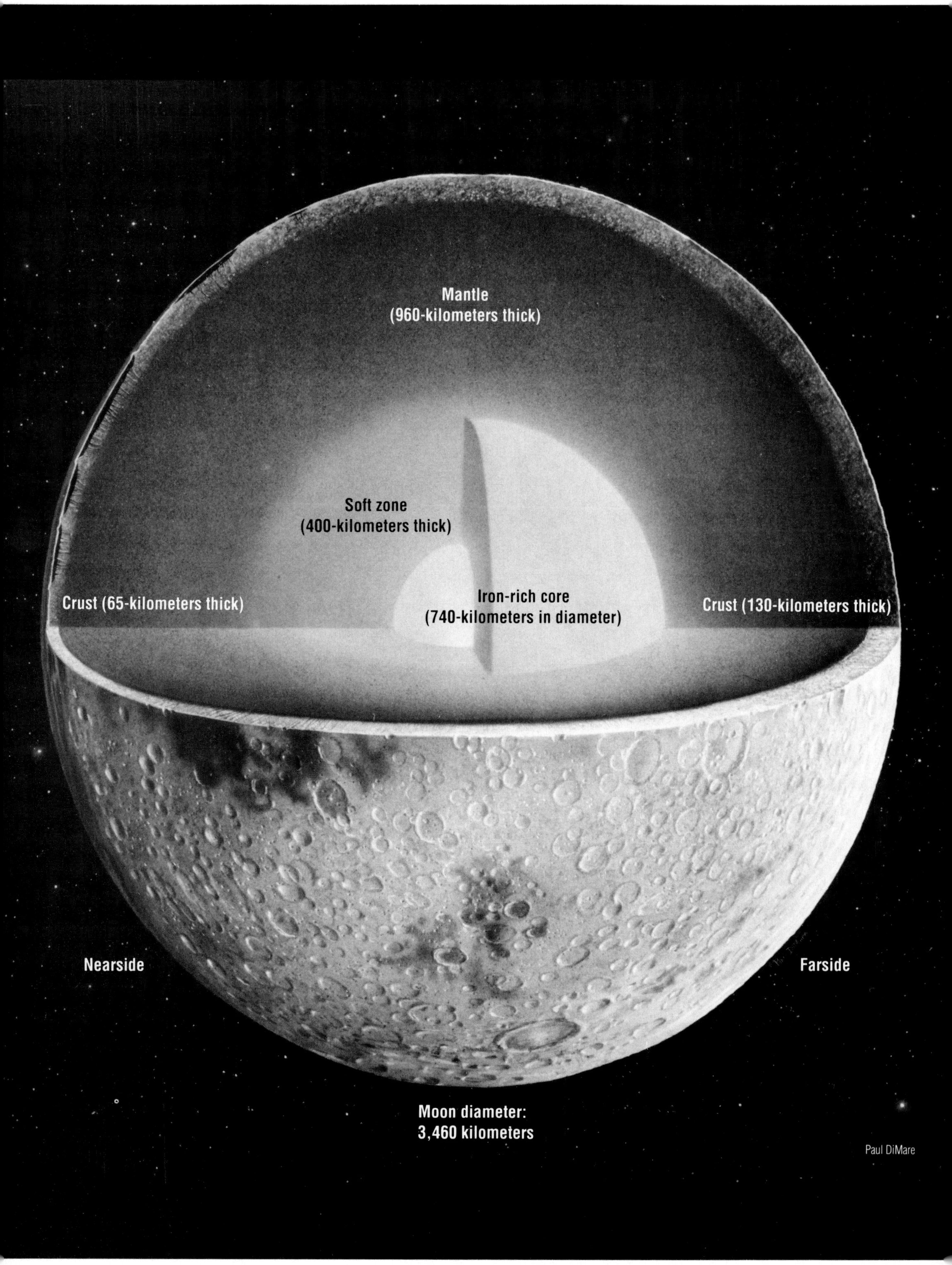

Mantle
(960-kilometers thick)

Soft zone
(400-kilometers thick)

Crust (65-kilometers thick)

Iron-rich core
(740-kilometers in diameter)

Crust (130-kilometers thick)

Nearside

Farside

Moon diameter:
3,460 kilometers

Paul DiMare

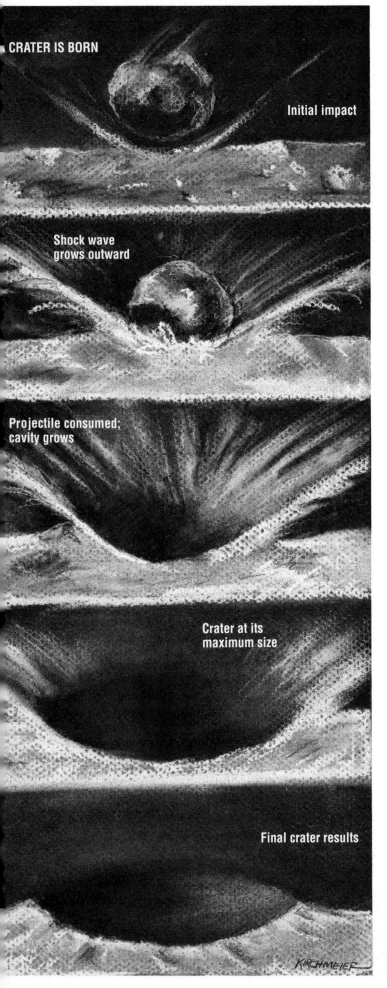

**CRATER IS BORN**

Initial impact

Shock wave grows outward

Projectile consumed; cavity grows

Crater at its maximum size

Final crater results

KIRCHMEIER

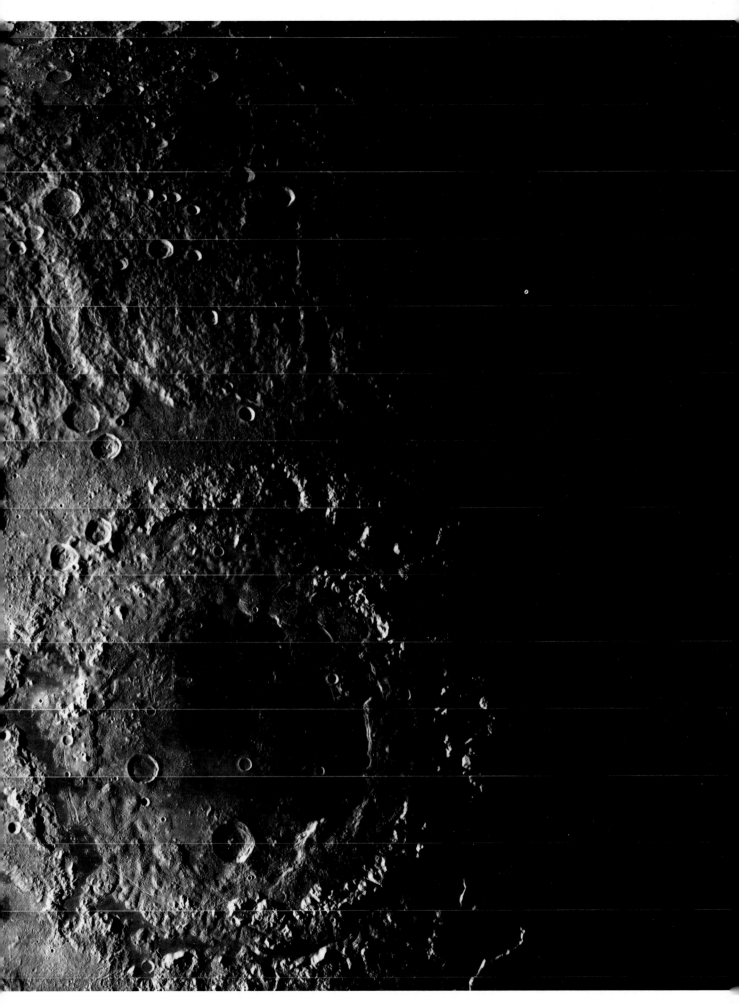

Iridum and Fracastorius. When a crater is located amidst an extensive lava field, it may be almost completely buried beneath lava flows. The feeble remnants poking through the lava are referred to as ghost craters. Many fine examples can be found in Mare Nubium.

Subsequent impacts can also modify existing craters. Huge amounts of finely divided ejecta will surge into the interiors of nearby craters, completely blanketing their floors. A fine example of this can be seen in the crater Ptolemaeus. Through a telescope, it is difficult to distinguish whether a specific crater's floor has been covered with lava or ejecta because the visual effect is similar.

The superimposing of one crater impact atop another can lead to many interesting effects. Examples exist of craters that have formed exactly on top of a central peak, making it look like a volcano with a vent hole, as in Regiomontanus. Other impacts have completely demolished the central peak, leaving a bowl-crater in its stead. Examples include Timocharis and Hercules. Sometimes an impact lands on a larger crater's wall, giving it the appearance of a diamond ring under certain lighting conditions. The craters Gassendi and Thebit show this effect beautifully. Rarely, a meteorite will impact dead center within an existing crater, forming another crater nested within the first. The best of these double-concentric bullseye craters is designated Hesiodus A. Across the Moon's surface, a multitude of impacts on top of impacts has given rise to a variety of illusory effects.

The ideal crater has a distinct set of features. Craters formed by recent impacts, such as Copernicus, Aristoteles, and Bullialdus, will exhibit every characteristic. But most craters have been more or less ravaged over time, and so no longer have a pristine appearance. In the extreme, ancient craters like Deslandres are barely recognizable as even *being* craters. The anatomy of a crater begins with the wall surrounding it. The outline of the wall is rarely circular, but instead tends to be polygonal, reflecting the effects of faults and other weaknesses in the regional terrain. Craters located near the limb of the Moon look like ellipses, but this is only a perspective effect.

The inner crater wall almost always slumps, which creates a series of terraces and sometimes deposits enormous tongues of landslide material on the crater floor. Slumping gives rise to a sharply defined crater rim, which may tower 20,000 feet above the crater floor. Over time, the rim will be pelted by small meteorites and gradually become softer in appearance. In the most ancient mare basins, only partial arcs of the original wall still remain intact; these form the sixteen named mountain ranges on the Moon.

**Shaped like a giant bull's-eye, Mare Orientale straddles the Moon's western limb and is visible only in part from Earth (preceding pages). This impact was the last major basin-forming event in lunar history. NASA photo.**

**Before it was flooded with lava, Gassendi crater (above, south at right) may have resembled Copernicus (right). But with Gassendi only the rim and the highest central peaks escaped inundation from Mare Humorum. By comparison, Copernicus is a recent feature being formed about 810 million years ago. The impact that created it came at a time when the lavas of Oceanus Procellarum had already flowed and hardened. Jean Dragesco photos.**

A Lunar Volcano? Believing it changed shape, astronomers once thought Linné (above) might be an active volcano. Space photos instead revealed that it is a small, fresh impact crater ringed with a bright halo of ejecta. NASA photo.

Crater floors are generally flat and depressed relative to ground level. But they are not as deep as they seem to be when viewed under grazing sunlight. If we represented a large crater like Clavius by a 25-cent piece, the coin would be four times too thick! The floor of a crater consists of highly fractured bedrock covered with a layer of rubble and (usually) small amount of pooled, melted rock created during the moment of impact. Some ejecta falls back into the crater, forming a hummocky blanket. Occasionally small domes and other evidence of local volcanism can be detected. Larger craters usually have small groupings of three to five central peaks rising 3,000 feet or more above the crater floor. Moderate-sized craters normally have a single central peak.

Surrounding each crater is an ejecta blanket. This is usually thick and continues out to one crater diameter, then grades to a thinner partial layer of debris. The ejecta blanket often displays radial patterns of hills and valleys, giving it a "ropy" appearance. Farther from the crater, you can see elongated secondary craters. These result from blocks of rock thrown out from the crater. Occasionally the secondary craters form an obvious herringbone pattern. In these cases the "V" of the herringbone always points back to the main crater.

The Moon does have volcanic craters, but they are small and insignificant compared to the great impact craters. Most are at the threshhold of detectability through a telescope and are often associated with rilles, which served as lava tubes for material emanating from the crater's vent. A fine example is the volcanic crater informally called the Cobra's Head, located at the source of Schröter's Valley.

**Lunar Ray Systems.** Fresh craters often have magnificent ray systems associated with them. These can extend for thousands of miles. As mentioned, rays consist of highly pulverized rock, called rock flour, which jetted from the crater immediately after impact. Rays run across old craters and even mountain ranges. They must be exceedingly thin, because none has ever been seen to cast a shadow. Rays are not always continuous white streaks but contain much structure and evidence of secondary impacts caused by blocks of material that travelled along with the rock flour. (Remember that in a vacuum both dust and rocks will fly equally unimpeded.)

Lunar enigmas still exist: for example, we cannot explain why some small craters have given rise to rays extending thousands of miles, while much larger craters have short ray systems — nor why the shape of the rays should vary so dramatically from site to site. These differences in form are most likely related to the physical structure of the impact area. Ray systems are generally short-lived, fading away after several hundred million years as the pulverized rock is discolored by solar wind ions and is mixed with local lunar soil by the gardening action of micrometeorites.

**Lunar Seas.** Searchers after lunar volcanos wasted their time hunting among the craters. They should have been looking at the lunar maria.

Volcanism on the Moon has taken a different form than that observed on Earth and Mars and is perhaps more like that on Venus. Volcanos on Earth and Mars are great constructs with high elevations. Examples are shield volcanos, such as Mauna Loa on Hawaii and Olympus Mons on Mars, and the strato-volcanos of Italy: Mts. Vesuvius and Etna. But the chemical composition of lunar basalts is such that they are almost as fluid as water. In consequence, lunar volcanism consists mainly of basalts flowing downhill and filling the lowest parts of the surrounding terrain.

In this way, the great impact basins were filled with thin sheet after thin sheet of lava. Photographs taken from lunar orbit clearly show tongues of lava flow fronts stretching hundreds of kilometers. Where the flows end, the elevation of the leading edge is so low that the shadow it casts even under the grazing sunlight is too small to be detected from Earth. The filling of the lunar basins proceeded slowly over long geological periods. In the case of Mare Australe, we apparently have an example of a mare arrested in mid-formation. The generous supply of basalts seems to have ended before enough lava had pooled to completely flood the pre-existing craters. What we observe, therefore, is a large group of flooded craters and flooded intercrater plains. And at the other extreme, seas like Mare Crisium present a smooth, almost featureless expanse of frozen lava.

It is not surprising to discover that since the seas represent the focus of lunar volcanism, they should also be the area in which we can observe the majority of volcanic and tectonic features of the Moon, namely the wrinkle ridges, rilles, and domes. Not a single one of these features can be found deep within the lunar highlands, although a fair number can be seen in highlands that lie immediately adjacent to the shore of a sea. The dark lavas of the lunar seas are much denser than the bright crustal materials they overlie. Wherever volcanism spewed out enough lava, its weight compressed the crust, which sought to reach a balance through a process called isostasy. Isostasy simply means that heavy materials sink until they reach a depth where the upward pressure is enough to float them; light materials likewise rise until they reach their balance point. Many lunar tectonic features are the result of isostatic adjustments.

**Wrinkle Ridges.** The largest tectonic features are the wrinkle ridges. These are long, low hills superimposed on the otherwise smooth surfaces of every one of the lunar seas. Most extend for hundreds of kilometers. Because they are only a few hundred feet high along most of their length, wrinkle ridges are best observed immediately after local sunrise or just before local sunset. When sunlight falls more steeply on them, they quickly become hard to detect.

Most wrinkle ridges extend in graceful arcs, and many seem to have a braided texture. They tend to follow the outlines of older craters and basins submerged beneath the lavas. There are several examples of lunar bays where the missing part of the crater wall has been replaced by a wrinkle ridge. (Letronne in the Mare Humorum region is a good example.) But others follow no particular plan, as in the case of Sinus Iridum, where a series of wrinkle ridges look like waves moving towards the shoreline and do not relate to any structural feature in the vicinity.

Mare wrinkle ridges are probably the least understood of all the lunar features. No less than six different hypotheses for their formation fit current observational data! One holds that they are simply folds that developed in the lava crust as it cooled and contracted. A second possibility is that they mark the locations of crustal faults through which lava welled to the surface. The ridges would then be the solidified final squeeze-out of molten lava. The truth may well be that mare wrinkle ridges have no single origin.

**Rilles.** Occasionally called clefts in older Moon writings, lunar rilles come in three basic forms. *Sinuous rilles,*

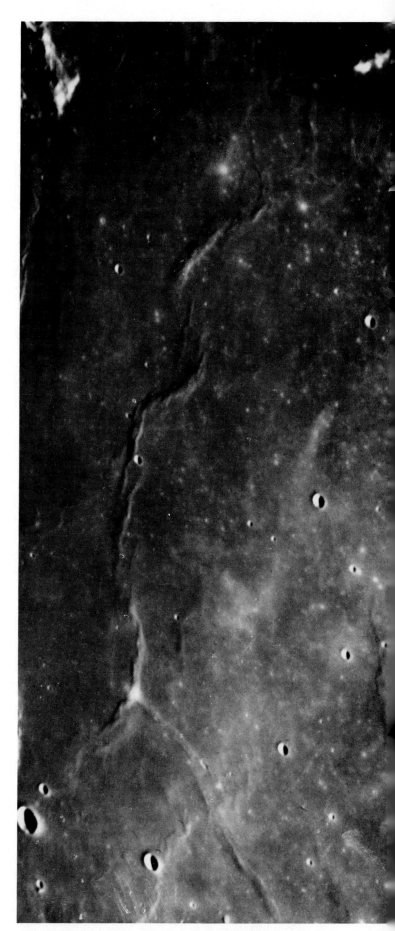

Like a rope made of rock, the Serpentine Ridge snakes for hundreds of kilometers across eastern Mare Serenitatis. Wrinkle ridges like this may form when the mare basin buckles under the weight of the lava filling it. Lick Observatory photo.

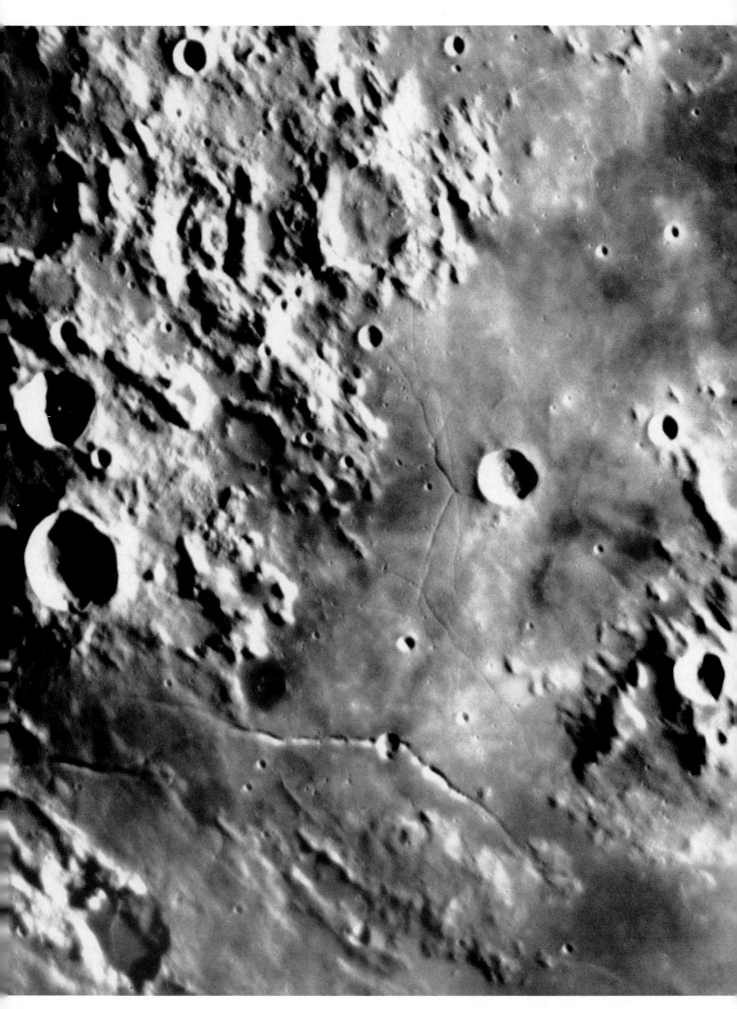

such as Rima Hadley visited by Apollo 15 at the base of the Apennine Mountains, are purely volcanic in nature. They represent collapsed tubes that carried molten lava from their source vents downslope to the lower-lying basin floors. Since the lunar lavas flowed freely, it is not surprising that sinuous rilles are river-like in appearance and have bends and elbows. In many instances, the source vent crater can be seen at one end of these rilles.

*Straight rilles* formed as a result of faults in the crust. These may have occurred as the result of basin-forming impacts, as for example Rima Sirsalis, or may represent structural weaknesses in narrow strips of lunar highland located between adjacent seas, as is probable for Rima Ariadaeus. Straight rilles cut across all terrain, including crater walls and floors, undiverging from their path. Simple faults result in relatively straight rilles with V- or U-shaped profiles. Frequently, two closely spaced parallel faults will cause the terrain in between to drop, resulting in a graben, a wide flat-floored rille.

*Arcuate rilles* are a variation on the straight rille. These are found near the perimeters of the lunar seas, and as their name implies, arcuate rilles curve in gentle arcs around these mare shorelines. They are purely tectonic, having formed as the dense basalts of the lunar maria sank under their own weight and bent the lava sheets around the basin perimeters until they cracked. Often two or more parallel arcuate rilles will form, creating an *en echelon* arrangement, as with the Hippalus rilles of Mare Humorum. Most arcuate rilles tend to be grabens.

All but the largest rilles are less than two kilometers wide. Their length and strong black appearance at sunrise or sunset combine to make them visible as fine, web-like lines on the lunar surface. Under very high magnifications, many display interesting detail, such as small wall slumps and twisted offsets. Generally, the best time to ob-serve rilles is about one day after local sunrise or before sunset — that is, when the terminator is about 12° of longitude away from the rille. Most disappear from view under higher lighting. Because lighting plays so critical a role in observing rilles, it's no surprise that those which trend north-south (crosswise to the direction of sunlight) are far easier to observe than those oriented east-west.

**Domes.** The last kind of formation associated with lunar seas are the domes scattered at random on or near the borders of the maria. The most common type of lunar dome has a low elevation and the profile of a flattened lens. Domes are usually featureless blisters on the Moon's surface, although many exhibit a rimless summit crater. Such low-relief domes have a strong resemblance to terrestrial shield volcanos, and this is most likely what they are. Shield volcanos are low, broad structures built up in thin layers like some kind of droopy rock torte. A particularly fine example is the dome Kies Pi, located in Mare Nubium. In a small number of cases, we can observe volcanoes where the summit craters have been breached by lavas, which then flowed down the side of the dome, creating a small rille.

However, other domes are steeper and more hill like. These often show some evidence of surface detail. This type, two of which are located near the crater Arago in Mare Tranquillitatis, do not have summit pits, which suggests they may have a different origin. Perhaps this type represents the lunar equivalent of terrestrial lava fountains, cinder cones, or may be localized uplifts caused by the presence of underground magma chambers.

The recognition of lunar domes as a distinct formation is a recent development. Indeed, the cataloging of domes was not carried out until the mid-20th century. Most domes have a diameter ranging up to 15 kilometers and even the larger ones may only stand 200 meters tall. This means their slopes incline at only a few degrees, and the typical dome is only visible in the grazing light of a lunar dawn or sunset. They can be challenging to see, and some remain visible for only an hour or two during each lunar month (although the more prominent specimens can be observed for somewhat longer).

**Straight cracks and twisted lines, rilles show where lava has been altered. Straight rilles like near Hyginus (lower center) are often due to faulting; meandering rilles next to Triesnecker (center) may show lava flow patterns. Gerard Therin photo.**

# 3. A Guide to Lunar Features

This section takes you on a guided tour of the Moon. In the pages that follow, I will tell you much about the major features of the Moon, including an outline of what to look for. Unlike many previous lunar guidebooks, I have also discussed the geology of formations, together with relevant historical notes. This is to give you the background you need to intelligently explore a region.

I have divided the Moon into fourteen regions, in most cases naming the region after the major sea that forms its focal point. Other regions are named for their most prominent feature. In each region simple directions will enable you to hop from one formation to the next. You will need a good quality map of the Moon (see Bibliography) to help locate these formations using my directions. Where a circular sea is involved, I have often resorted to locating features along the perimeter by visualizing the sea as a clock face. In each case, I have used the convention of designating due north as noon and (lunar) due east as 3 o'clock. This may require you to make left-to-right or up-and-down adjustments depending on your type of telescope and accessories. (Star diagonals, for example, give you a mirror image, reversed left for right.)

For each feature mentioned, I will point out how to find the smaller details of interest but which may not be apparent at first glance. Except where noted, every observational detail I mention will be readily visible through a good 4-inch refractor or 6-inch reflector on a night of good seeing. In all, I have covered approximately 250 lunar features.

The observer who works completely through the guide and views all the objects will have touched upon a great deal of what the Moon has to offer. In the process, the enormous geological forces that have come to play in shaping the lunar surface will be revealed firsthand. Perhaps most important, you will have had the opportunity to enjoy the very same formations that have thrilled lunar observers since Galileo first lifted his telescope to the Moon.

## MARE CRISIUM

Each month when reappearance of the crescent Moon after sunset announces a temporary end to deep-sky observing, the first lunar formation we are drawn to is the dark oval of **Mare Crisium,** the Sea of Crises. This sea, which measures about 500 kilometers across, is the best preserved of all the lunar maria and is still completely encircled with a ring of mountains. (Although Crisium appears oval in a north-south direction, it is actually oval in an east-west direction; the foreshortened perspective we view it at is to blame.) In Mare Crisium we can see that the lunar seas are products of immense impacts. With some other cases the evidence for impact has been long obliterated, as in Mare Tranquillitatis not far to the southwest, but Crisium still displays its heritage.

Although it is 3.85 billion years old, Mare Crisium is so fresh (in lunar terms) that its floor is an almost blemish-free expanse of frozen lava. Only four small craters, all clustered near the western rim, are found on its floor. The largest is **Yerkes**, which is only 36 kilometers across. This is a flooded crater with some curious appendages to the north and south that probably represent the remains of some adjacent craters. **Lick**, another flooded crater, lies just to the south. Due east of Yerkes is the bowl-shaped crater **Picard**, only 23 kilometers in diameter. And to the north of these is another bowl-crater called **Peirce**. Half a dozen small craterlets can also be found in this region, but you would be hard pressed to find an equal number of still-smaller craters on the rest of Mare Crisium, even on a fine night. The northern part of the sea shows a small number of faint ghost craters also.

Examining Yerkes and Lick helps illustrate how flooded craters form. The impact that created the basin which Mare Crisium fills would have totally obliterated any pre-existing craters. Both craters therefore postdate the basin-forming impact, although by how much is uncertain. The lava that oozed out to fill the basin encroached on the new craters and gradually ate away at their floors and walls, leaving the remnants we see today. Since the basin excavated during the Crisium impact was deepest in the middle, small craters formed there have been totally drowned by lava flows. This is why in general we see flooded craters only near the margins of Mare Crisium and the other lunar seas.

Several interesting formations can be found near the western rim of Mare Crisium. Almost due west is the prominent 28-km crater **Proclus**, the center of a bright system of rays. Curiously, the rays extend out from Proclus in all directions except to the southwest, where there is a 150°-wide zone of exclusion in the region of a small mare area, **Palus Somni**. There is no ready explanation for this effect. Proclus is a fresh crater, and is the second brightest spot on the Moon (after Aristarchus crater in Oceanus Procellarum). As the Sun rises higher over Proclus, the crater interior brightens rapidly, making it difficult to see clearly details within it.

Close by Proclus, on the western rim of Mare Crisium are two capes, **Promontory Olivium** and **Promontory Lavinium**, which present an observational mystery. These stand halfway between Proclus and the flooded crater Yerkes. In the moments just after sunrise or before sunset here, these two capes seem to be connected by a bridge. It looks so artificial that there are routine reports from neophyte observers convinced that someone is constructing a highway on the Moon. In reality, it is just the jumbled remnant wall of an ancient crater, but the illusion is powerful — see for yourself!

To the north of Proclus is the crater-pair **Macrobius** and **Tisserand**. Macrobius, the larger of the two at 64 kilometers, is the western member of the pair and has an in-

**One eye of "The Man in the Moon," Mare Crisium is visible about three nights after New Moon. Around its shores lie impact craters almost obliterated by the lava floods that filled the Crisium basin to make the mare. Christian Arsidi photo.**

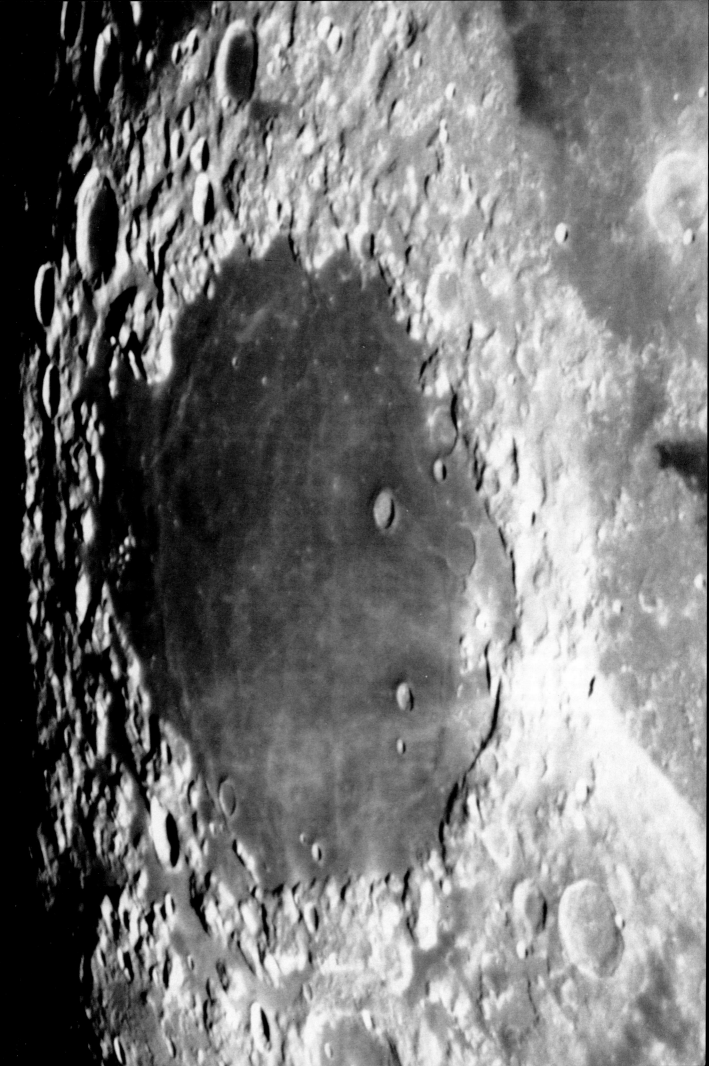

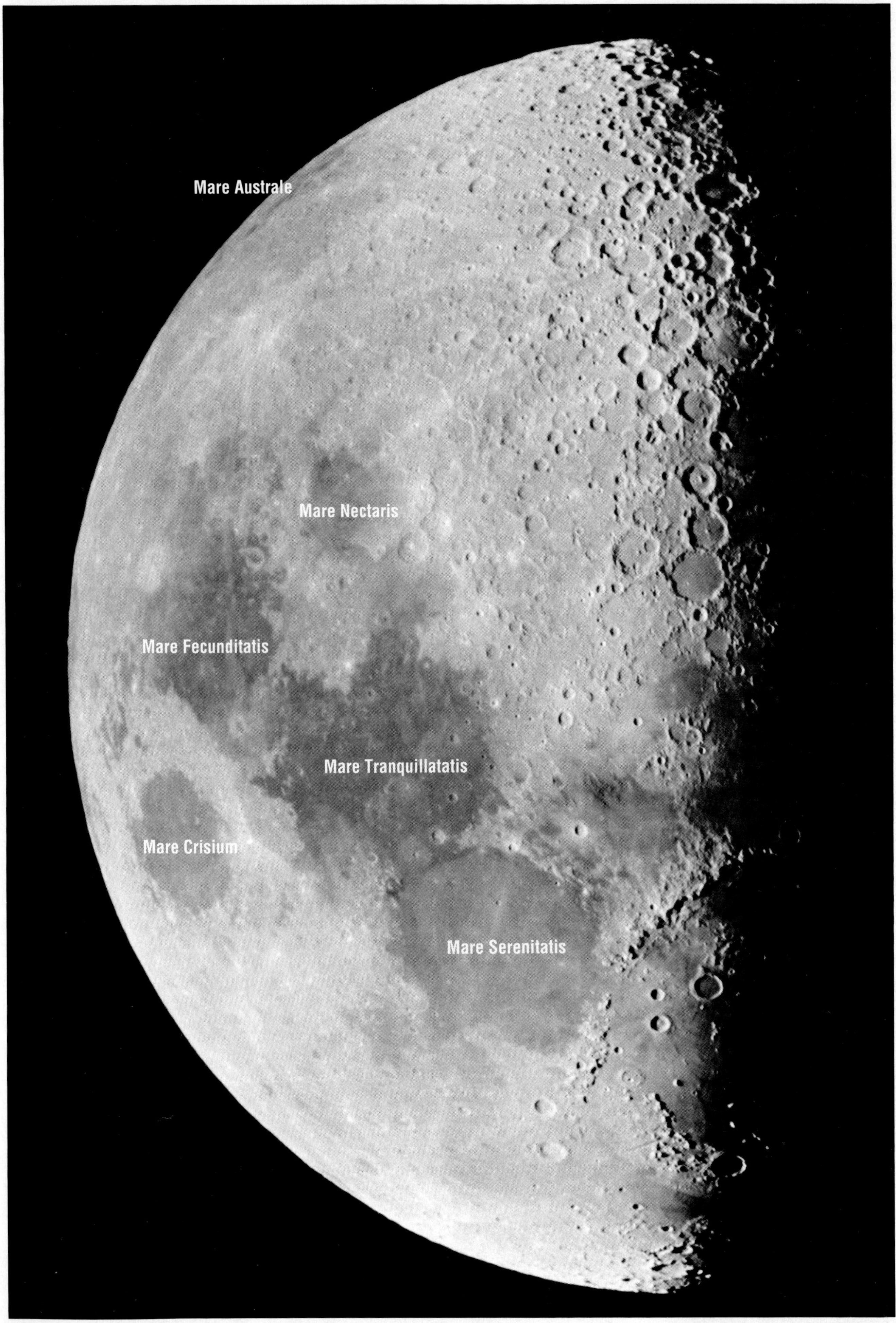

Mare Australe

Mare Nectaris

Mare Fecunditatis

Mare Tranquillatatis

Mare Crisium

Mare Serenitatis

Six lava "Seas" are visible at First Quarter Moon. Three of them (Crisium, Serenitatis, and Nectaris) are lava flows that fill the scars left when meteorites struck. Above: Looking like a comet, the twin tails of ejecta from Messier and Messier A suggest a low-angle impact for the origin of the crater pair. Photos by Brian Kimball (left) and Richard J. Wessling (above).

teresting central mountain formation. These chance pairings of craters are common across the face of the Moon. Although the individual craters in the pairs are unrelated in a geological sense, their closeness to each other makes them more attractive and in many cases provides for an interesting study in contrasts.

Moving further counterclockwise around the rim of Mare Crisium, we next come upon the large and prominent crater **Cleomedes**, 125 kilometers in diameter. On the smooth flooded floor of this crater can be found two perfectly formed bowl-craters, a tiny central massif, and a nice rille which cuts across the northern half of the floor.

## MARE AUSTRALE

The extreme southeastern limb of the Moon contains numerous dark patches which collectively make up **Mare Australe**, the Southern Sea. This formation stretches well over onto the farside of the Moon, and it is therefore best seen when favorable librations swing the region toward us. Also, Mare Australe should be examined just after First Quarter Moon, when the Sun stands at local noon high over this formation, and the dark areas contrast most strongly with the highlands around them.

Of course, photographs of Mare Australe taken from lunar orbit show it far better than the edge-on perspective we are treated to from Earth. They suggest that this is a mare whose formation was arrested mid-stream. Mare Australe's true shape is circular and it is only slightly smaller than Mare Crisium. But only about half its surface area is covered with lava. The dark patches we observe are individual craters and small groups of them whose floors were flooded with lava. It appears that the volume of lava was not sufficient to completely cover the crater rims and higher terrain. Thus the Australe basin was never

totally filled, as is the case with the nearside mare we are more familiar with.

The most obvious of the flooded craters comprising Mare Australe is 140-km **Lyot**, located close to the center of the mare basin adjacent to the Moon's limb. More than 140 kilometers wide, Lyot's floor is dark, making its detection easy. However, located as it is almost on the Moon's limb, this nearly circular crater is extremely foreshortened into a tight ellipse by the perspective of the view.

## MARE FECUNDITATIS

To the south of Mare Crisium is the irregularly shaped **Mare Fecunditatis**, the Sea of Fertility. On its floor are a number of wrinkle ridges, a smattering of small bowl-shaped craters, and the showcase crater pair **Messier** and Messier A, the "comet craters" (see below). The perimeter of this 850-km mare is sprinkled with interesting formations. On the eastern border is the magnificent crater **Langrenus**, 130 kilometers across. Larger than its cousin crater Copernicus far to the west, Langrenus has a prominent system of terraces on its inner walls and two fine central peaks. Like Copernicus, Langrenus has a conspicuous ray system but being older, the rays have faded considerably. Were Langrenus located closer to the center of the lunar disk it would certainly be far more widely observed than it is. Due east of Langrenus, lying astride the lunar limb, is the more circular sea **Mare Smythii**, Smyth's Sea. Depending on libration, it is alternately seen as a thin, dark strip along the edge of the Moon or as an ellipse whose outer basin walls can be traced fully around its circumference.

South of Langrenus is **Vendelinus**, a severely battered old crater 145 kilometers in diameter. Continuing south along the meridian, the crater **Petavius** completes this lineup of large craters. Petavius is still larger (177 km) than Vendelinus and is in a much better state of repair (see page 36). It has a prominent rille on its floor. Also in this region is the crater chain **Vallis Snellius**. Still further to the south are the craters **Furnerius** (125 km) and **Stevinus** (75 km), noteworthy because two bright craterlets associated with them will develop the largest ray systems in the southeast quadrant of the Moon as the Sun rises higher.

Directly across from Langrenus on the western shore of Mare Fecunditatis are the flooded craters **Goclenius** (60 km) and **Gutenberg** (70 km), as well as an assortment of a half dozen or so other flooded craters to their south. A system of parallel rilles emanate from Goclenius, fanning out to the northwest as they run for over 150 kilometers. Parallel rilles can also be found cutting across the floors of both Gutenberg and Goclenius. These demonstrate that this entire rille system was formed after the craters were created by impact, then eroded over time, and finally were flooded by the Fecunditatis lava flows.

On the north end of Mare Fecunditatis, close to where it merges into Mare Tranquillitatis, is the delapidated crater **Taruntius**. This shallow 55-km crater has concentric walls and a relatively well-preserved central peak. The nested walls are subject to interesting and unpredictable variations in albedo and highlights as the Sun rises over the crater.

### Messier

The western plains of Mare Fecunditatis contain the well-known and much studied pair of craters, **Messier** (the more eastern of the two) and **Messier A**. (Older maps and lunar books sometimes refer to Messier A as W. H. Pickering, a name that is no longer used.) Though small, these two craters are highly distinctive and easy to find because of the bright, slightly diverging rays that radiate westward from these craters. The rays give the formation the unmistakable look of a comet caught in flight across the face of the Moon. No other lunar feature looks quite like this.

The Messier craters become visible the third night after New Moon. Messier is distinctly elongated (8 by 11 kilometers), while Messier A appears more circular (11 by 13 km). The floor in both is noticeably darker than the inside walls. Messier A is actually a double crater with a rounder eastern half that looks brighter. It can look triangular under grazing illumination or when the seeing isn't steady. If you view the craters over three or four consecutive nights, the play of shadows in their interiors will alter their apparent shapes considerably. They are fascinating because the longer you look at them, the more difficult it will seem to be to account for their appearance.

Many theories have been proposed to explain the formation of this odd pair. Because none of the theories is entirely satisfactory, and because of the strange and varying appearance the Messier craters present, we can conclude that whatever the cause, the outcome was one of those occasional freaks of nature. Nonetheless, this formation is so intriguing that it was specifically targetted for high-resolution photography during the Lunar Orbiter and Apollo missions. Careful analysis of the true shapes of each crater and of the unusual ejecta patterns suggests that the most likely mechanism of formation involves a grazing impact of a small, westward traveling meteor.

This meteor may have broken into two pieces prior to or at impact, or it may have skipped across the lunar surface after forming Messier to impact one more time and form Messier A. Lateral lobes of ejecta and the presence of the famous twin "comet tail" indicate the meteor hit the surface at an angle of about 5°, causing the elongated shape of these twin craters.

### Petavius and Vallis Snellius

At the southeast end of Mare Fecunditatis lies the unique crater **Petavius**. Measuring 177 kilometers in diameter, it is close enough to the lunar limb to appear elongated due to foreshortening. Actually, it is nearly circular. The feature that immediately attracts attention is a very wide rille running radially from the crater's massive central mountain complex out to its inner rim. This rille presents itself as a white line, however observation at higher powers will bring out structural details. Also, careful examination will reveal that this rille intersects a second rille running parallel to the inner base of Petavius' wall. Larger telescopes may disclose the presence of two other radial rilles spaced at 120° intervals around the central peaks. However, these are fine and somewhat difficult objects to see.

These rilles formed when the floor of Petavius, which apparently was flooded with lava (common in large craters located adjacent to a mare), subsequently cooled, and contracted. When the Sun is shining higher over Petavius at First Quarter Moon, you will spot several dark patches on its floor. This confirms the presence of volcanic lava flows, since dark colors on the Moon are usually associated with iron-rich basalts. This was a major finding of the Apollo 17 mission, which explored the dark-floored Littrow Valley to the north of Petavius and confirmed

what lunar investigators had deduced from telescopic observations. However, Apollo 17 also provided a surprise. Investigators had long thought the dark patches represented the freshest lava flows (hence the selection of the Apollo 17 landing site). As it turns out, the opposite holds true, and these appear to represent the most ancient volcanic activity.

Butted up against the eastern wall of Petavius is a highly elongated structure called **Vallis Palitzsch**. Reminiscent of the much larger Schiller (located in the southwestern quadrant), it is most likely a compound crater — that is, multiple adjacent impacts that coalesced into a single formation. Perhaps the impact that formed Petavius crunched Vallis Palitzsch even further. This is an area that was poorly imaged by the unmanned Lunar Orbiter probes and certainly deserves more detailed investigation.

Immediately southwest of Petavius is the relatively fresh crater **Snellius**, only half its size and also containing central peaks. Snellius interrupts a long and somewhat narrow crater chain called **Vallis Snellius** which runs several hundred kilometers. Note that it is oriented radial to Mare Nectaris, located to the northwest. This provides a clue to its origin. During the massive impact that created the Nectaris basin, huge gouts of material were ejected in all directions. Vallis Snellius is a line of secondary impacts from one squirt of gigantic blocks of rock. Because of the extreme age of the Nectaris Basin, most of the other similar features have long been obliterated. But if you scan this area under grazing light, you can pick up other examples. The largest of these is Vallis Rheita, located not far to the southwest (see page 39). A general survey of the entire region, especially as the terminator passes across it shortly after Full Moon, will uncover hints and suggestions of numerous unnamed radial valleys.

### MARE NECTARIS

With a diameter of about 300 kilometers, **Mare Nectaris** is one of the smallest of the circular lunar seas. The basin that the dark lavas presently fill is also one of the oldest that survive, having an age of 3.92 billion years. Located just to the southwest of Mare Fecunditatis, Mare Nectaris has a floor that is mostly featureless, except for a system of wrinkle ridges near the western shoreline. Close to the rim are a few flooded ghost craters, most notably 45-km **Daguerre** to the north. But the area immediately surrounding Mare Nectaris has many formations worth observing. On the western shore is one of the most attractive crater groupings on the Moon, consisting of **Theophilus**, **Cyrillus**, and **Catharina** Matched in size, they are different in age and provide an interesting geological sequence, as described on page 39. At sunrise six days after New Moon, this is a striking multiple formation dominating the global view of the Moon.

East of Theophilus on the mare is the irregular 28-km crater Mädler, whose floor is crossed by a ridge. A splash of bright material extends across Mare Nectaris to the northeast. Moving clockwise around the rim of Mare Nectaris, the small crater **Capella** (45 km) is found at the north end. A curious crater chain passes across this crater, looking at sunrise like a dagger plunged through Capella's

**Central peaks and terraced walls are highlights of Langrenus (bottom right) and Petavius (center left), situated on the terminator in this photo. Notice also the rille running across the floor of Petavius from central peak. Jean Dragesco photo.**

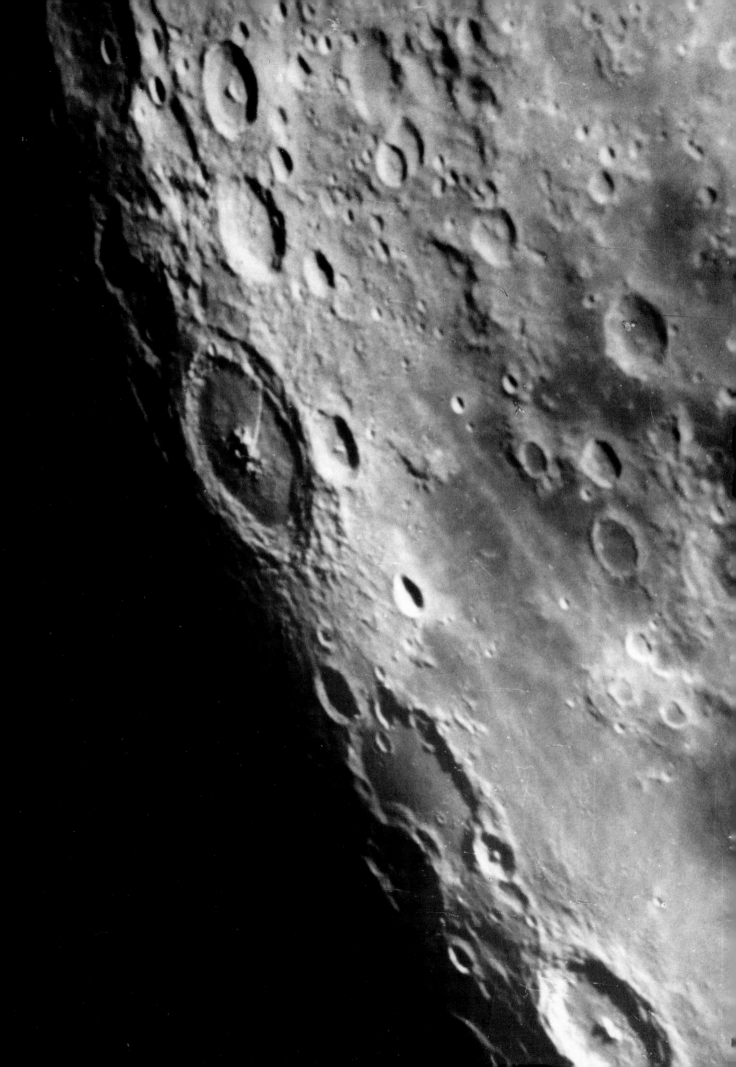

heart. The eastern edge of Mare Nectaris is bounded by the **Montes Pyrenaeus**, and just inside the rim is the 33-km crater **Bohnenberger**, which has an uneven floor rumpled with many hills. The southern side of Mare Nectaris is dominated by the partially destroyed crater **Fracastorius** (124 km), whose northern wall is largely missing. Where it once stood all that remains are a few scattered hills. A miniscule remnant of the central peak of Fracastorius can be seen poking through the smooth lava-clad floor of this crater.

In the highlands to the southeast of Mare Nectaris is the fascinating and unique formation **Vallis Rheita**, looking like a dark gash on the Moon's surface at sunrise or sunset. This formation is discussed on page 40. As noted earlier, this valley, together with Vallis Snellius and several other unnamed valleys in the region are the result of impacts by ejecta tossed out of the Nectaris basin when it was formed. Southwest of Vallis Rheita is the huge (190 kilometers in diameter) and ancient crater **Janssen**. Greatly demolished after billions of years of being battered by meteor impacts, it is hard to recognize it as a crater except at sunrise or sunset. Of interest is the unusually wide rille arcing across Janssen's floor.

Returning our attention to Mare Nectaris, one other interesting formation is associated with it. This is the **Rupes Altai**, or Altai scarp, which is concentric to Mare Nectaris outside its rim to the southwest. Brightly illuminated at sunrise six days after New Moon, the scarp represents the one clear remnant of the outer rim of the Nectaris basin. In its pristine state, the Nectaris basin must have looked like a multi-ring bullseye. Subsequent impacts have destroyed all but this one graceful arcing section. The Altai scarp is cut by a prominent set of four cross-faults which are easily spotted. At the southeast tip of the scarp is one the the the most beautiful impact craters on the Moon, Piccolomini, 90 kilometers in diameter, with terraced inner walls and a huge central mountain peak.

### Theophilus, Cyrillus, and Catharina

The western shore of Mare Nectaris is defined by the trio of craters Theophilus, Cyrillus, and Catharina. Matched in size at about 100 kilometers' diameter, these present the observer with side-by-side examples of craters with vastly different ages. This difference results in each crater having individual characteristics which make it distinctly different from the others.

**Theophilus**, the northernmost of the group, is youngest and displays all the characteristics of youthful craters. (In this it joins Copernicus, Langrenus, and Aristoteles, to name but three.) It has well-preserved and sharply defined rim walls with a highly developed system of terraces on the inside and an ejecta blanket outside. Under high sunlight, several weak rays can be seen to emanate from it. The terraces are extremely complex and well worth study. The southwest portion, which abuts Cyrillus, has undergone particularly extensive slumping. Theophilus has a group of three large central peaks which cast ever-changing shadows. The floor is smooth in the north, but contains fine detail elsewhere.

Moving southward, we next come to **Cyrillus**. It's easy to tell that this crater predates Theophilus. In addition to a lack of sharpness in features characteristic of fresh craters, it is obvious that the Theophilus impact has destroyed part of the wall of Cyrillus. Like Theophilus, Cyrillus has a group of three central peaks, but these are puny in comparison. The floor of Cyrillus has been extensively modified with age and contains considerable detail. Just south of the central peaks is a wide curving rille. A 10-km bowl-crater, **Cyrillus A**, has impacted just inside the western wall. Lastly, no evidence of ejecta can be detected in the surroundings. All these observations demonstrate that

**Lit by a Rising Sun, the Altai Scarp draws a bright line as it curves around Mare Nectaris. Inside the scarp lie the large craters Theophilus, Cyrillus, and Catharina (bottom to top). Above: Janssen is an old and much ruined crater. Heavily modified by later impacts, little of its original floor remains. The southern highlands are so saturated with craters that each new impact will destroy the traces of older ones. Photos by Marc Duncan (left) and Christian Arsidi (above).**

Tranquility Base

Cyrillus is considerably older than Theophilus.

**Catharina** is located further south, slightly separate from Cyrillus. One look will convince you that this is an ancient crater indeed. Its walls have been severely battered. Four impacts, each in turn quite old, have obliterated the northeast portion of the wall. The largest of these, 45-km **Catharina P**, has itself been significantly degraded by the ravages of time and is now not much more than a ghost. No sign of the central peaks that must have once existed remain and traces of terraces on the inner wall are barely detectable. The floor of Catharina is smoother than that of Cyrillus. This is most likely the result of features being buried under a sheet of ejected debris from other nearby impacts.

Theophilus remains a prominent object throughout the lunar month. When the shadows cast by its rugged ramparts disappear, a bright ring representing pulverized rock emplaced on its outer rim makes spotting this crater easy. On the other hand, as the lunar morning progresses, decrepit Catharina soon becomes a difficult object, with Cyrillus following not far behind. Their shadows shorten quickly as the Sun rises, indicative of their lowered relief

thanks to the incessant pelting of the lunar surface by micrometeoroids. But once upon a time, even Catharina was every bit as magnificent as Theophilus is today.

### Vallis Rheita

**Vallis Rheita** (or the Rheita valley) becomes visible four days after New Moon. It lies southeast of Mare Nectaris in a jumbled highland area littered with ancient craters. The valley, which at local sunrise looks like a wide black gash, consists of a chain of craters with broken-down walls between them. One oblong hole overlaps the next to create a valley almost 200 kilometers long and about 25 kilometers wide, and which is perhaps a kilometer

**The first humans landed on the moon in Mare Tranquillitatis. Although nothing at Tranquillity Base is visible from Earth, it's still a thrill to locate the site of the first moonwalk. Right: In western Mare Tranquillitatis, the buried crater Lamont shows a ring of ridges lit by the setting Sun. A few domes also lie scattered among the fresh craters on the mare surface. Photos by Jiri Bunata (above) and Jean Dragesco (right).**

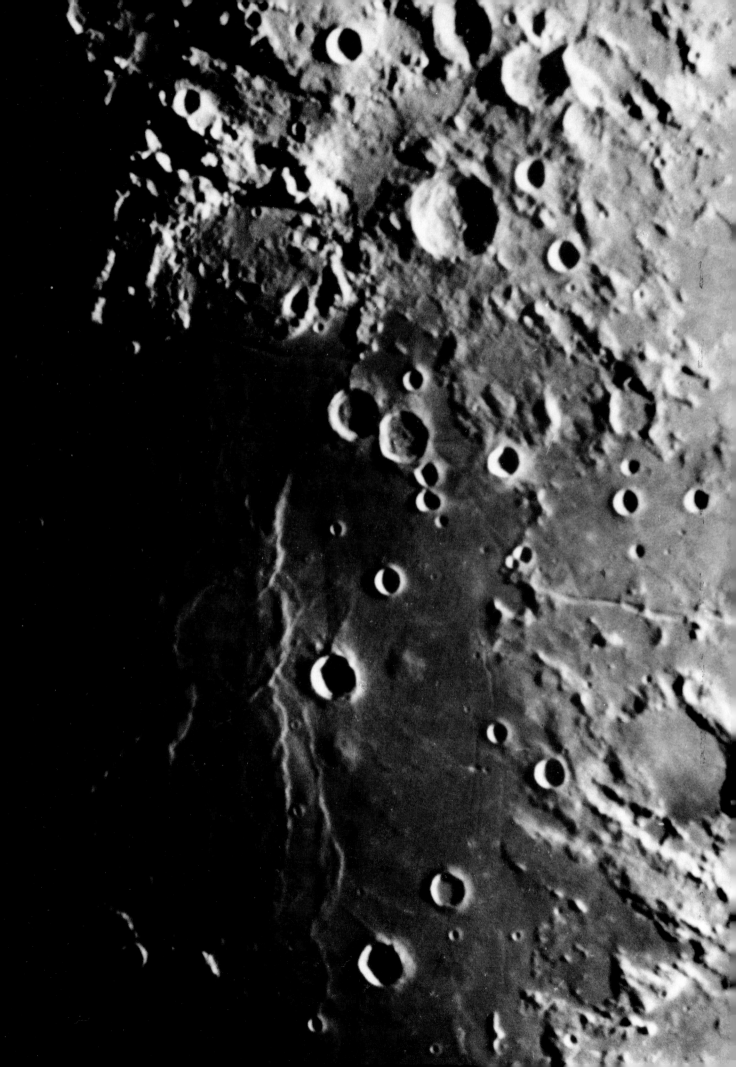

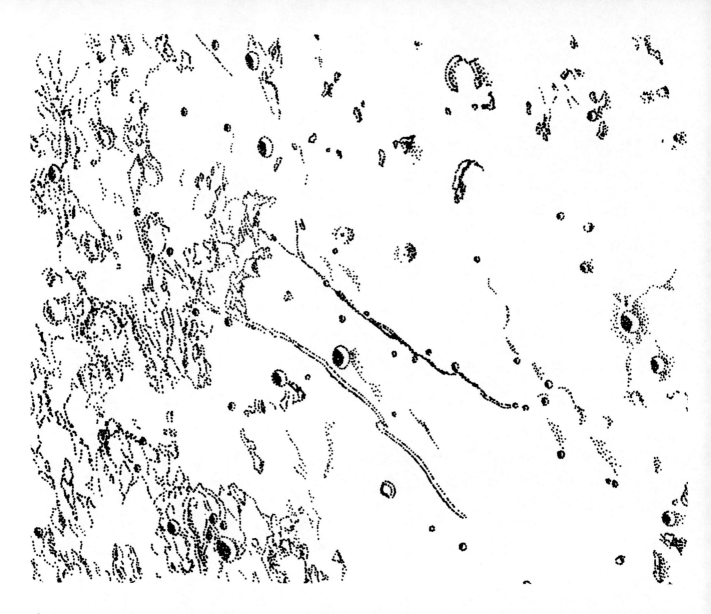

deep in spots. Also a poorly defined extension angles away from its southern end and continues for at least another 200 kilometers.

The Rheita valley begins on its north end adjacent to its namesake crater, **Rheita**. This 70-km crater is much younger than either the valley or most of its neighboring craters. It has some terraces and a well-developed central peak, making its identification easy in an area that is otherwise difficult to decipher. The valley itself is composed of seven or eight coalesced old craters. Its irregular sides and buckled floor are especially interesting under low-angle illumination, as occurs 4 to 5 days after New Moon and again 3 to 4 days after Full. In two spots close together and another to the north, you can see where rocky ridges cut across the floor, breaking its length. These are the last remnants of the old crater walls.

Although a few other similar formations exist on the visible side of the Moon, these are are located close to the limb. Examples are Vallis Snellius near the southeast corner and the Bouvard, Inghirami, and Baade valleys, located on the extreme southwest edge of the Moon. These are all difficult objects to observe. The Rheita valley, on the other hand, is very well placed and intriguing to explore, although it becomes relatively hard to see clearly when the Sun illuminates it from a high angle.

In the pre-Apollo days, when volcanism was believed to be the dominant force shaping the surface of the Moon,

it was believed that Vallis Rheita was a chain of volcanic calderas aligned along a massive fault radial to Mare Nectaris. However, overhead views of the lunar farside obtained by Lunar Orbiter show that Mare Orientale and the enormous crater-basin Schrödinger both possess radial valleys which are larger and better preserved. Examination of these provides confirming evidence that these formations result from secondary impacts during big basin-forming events.

We can now state that Vallis Rheita was formed in the immediate aftermath of the giant impact that created the Nectaris basin to the northwest. Jets of huge rocky blocks were thrown out during this impact and these travelled as a group to impact in a linear arrangement. It is not easy to determine this mode of formation by viewing just the rather degraded Rheita valley. However, the Schrödinger event, being of more recent vintage, has preserved the evidence far better. In it scientists find pristine crater rims

**Variations in tone hint at differences in age and composition for the lavas of Mare Serenitatis. The mare also contains a large ray pointing toward the distant southern crater Tycho, a giant wrinkle ridge, and the once-enigmatic crater Linné. Above: The small crater Cauchy lies between two fault scarps in eastern Mare Tranquillitatis. Several domes also stand in the region. Lick Observatory photo (right), drawing by Patti L. Keipe (above).**

Linné

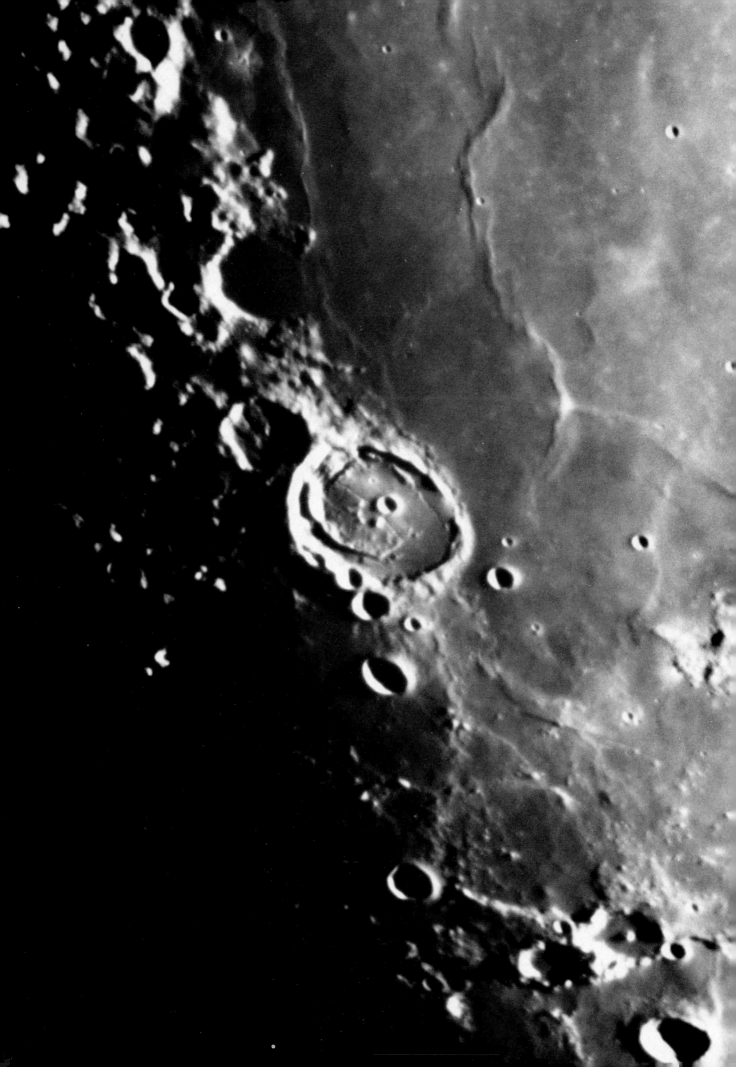

and the telltale herringbone ejecta patterns indicative of secondary impacts. Unfortunately, we can observe the Schrödinger valleys only in photographs released by NASA, as this formation lies just around the lunar limb, coincidentally in the direction pointed to by the Rheita valley.

## MARE TRANQUILLITATIS

Mare Tranquillitatis, the Sea of Tranquillity, will be remembered forever as the site of mankind's first extraterrestrial excursion (photo on page 40). This broad, irregular sea is one of the oldest of the lunar maria and its shape, once circular, has been substantially altered by the later impacts that created Mare Serenitatis to the north and the Fecunditatis and Nectaris basins to the south. On the eastern side of Mare Tranquillitatis is a marvelous vista of domes, rilles and faults in the vicinity of the crater **Cauchy**. These are described on page 47.

The southwestern corner of this mare also holds a wealth of interesting objects. The obvious guidepost in this area is the matched pair of craters, **Ritter** and **Sabine**, which lie side by side. These craters have unusually textured floors apparently modified during the era of volcanism. **Tranquillity Base**, site of the Apollo 11 landing, is located in a smooth, seemingly featureless expanse of lava two Sabine-diameters due east of the southernmost

**Partly destroyed by floods of molten rock, Posidonius is a large crater that was born during the interval between the excavation of the Serenitatis basin and the lava flows that filled it. Above: Late afternoon Sun lights the rugged terrain around Aristoteles and Eudoxus, while farther east the ramparts of Bürg catch the last rays of the setting Sun on Lacus Mortis. Photos by Gérard Therin (left), Jean Dragesco (above).**

point of Sabine's rim. This area was also explored by two unmanned probes, Ranger 8 and Surveyor 5.

North of Tranquillity Base is the crater **Arago**, which is about the same size as Sabine but of more recent vintage. Arago is definitely distorted from circular and a ridge can be seen extending from the north end of Arago's wall to the center of the crater floor. Perhaps this is a freak form of an elongated central peak. In any case, the prime attraction of the Arago region are two large domes, one less than a crater-diameter due north, the other a crater-diameter due west. These are best seen shortly after the terminator passes through the area. Unlike most other lunar domes, which look like flat blisters and have smooth surfaces, these two (Arago Alpha and Beta) are more complicated and have a fair amount of structure easily seen through the eyepiece. By way of contrast, there is a group of four small typical blister-like domes northwest of Arago Alpha.

Immediately to the southeast of Arago is one of the strangest formations on the Moon, seen in its fullest detail only as the terminator passes across it. **Lamont** is an inconspicuous formation some 90 kilometers across, whose "walls" are outlined by wrinkle ridges. Lamont is the focus of a complex system of wrinkle ridges heading radially in all directions. The circular portion of this formation may mark the location of a now-buried crater, but there is no explanation for why this should have given rise to the scores of braided wrinkle ridges that seemingly emanate from it.

A system of rilles can be traced with difficulty all around the southeastern edge of Mare Tranquillitatis. Most are concentrated in the vicinity of Ritter and Sabine. The largest of these is **Rima Hypatia**, which travels east from Sabine along the southern shore of Mare Tranquillitatis.

## Cauchy

In the eastern portion of Mare Tranquillitatis near Palus Somni lies the quite undistinguished crater **Cauchy** (photo on page 42). It can be located at the western tip of an equilateral triangle formed with Proclus and Taruntius. Measuring a mere 12.5 kilometers in diameter, Cauchy is like scores of other small, relatively fresh craters. Its rim is sharply defined, its interior is like an bowl, and it is bright and easily spotted at Full Moon. Yet this otherwise run-of-the-mill crater is the focal point for one of the most diverse assemblages of geological formations on the Moon.

If you observe Cauchy about five days after New Moon, your eye will be drawn to a slender dark streak jutting out of the terminator and running eastwards just to the south of Cauchy. This is one of the rarest features on the Moon, a thrust fault. There are only two other similar structures on the Moon, Rupes Recta (the famous Straight Wall) located in Mare Nubium, and Rupes Liebig located on the western edge of Mare Humorum. Both of these are oriented along the north-south axis of the Moon, which means that they cast long shadows when the Sun is appropriately placed.

**Rupes Cauchy** on the other hand is oriented about 30° off the east-west axis of the Moon, so the shadow it casts is not nearly as large. Were it oriented north-south it would surely be one of the most prominent features in that region of the Moon. As it is, the dark shadow of Rupes Cauchy is easily spotted in small telescopes. In an 8-inch scope you can see that the fault turns into a delicate rille about two-thirds of the way along its 250 kilometer length. Three to four days after Full Moon the lighting is reversed and with the Sun shining more fully on the inclined wall of Rupes Cauchy, it appears as an even more slender bright streak.

Running parallel to Rupes Cauchy but on the northeast side of Cauchy crater is another excellent rille. This feature originates near the ruins of an unnamed crater and is just as long as the fault. Given their parallelism, it is likely that these two structures — Rupes Cauchy and **Rima Cauchy** — are geologically related. Finally, a little south of Rupes Cauchy are two excellent domes, named Omega Cauchy and Tau Cauchy. They are as fine as any you'll see elsewhere on the Moon. Visible under grazing light in small telescopes, large backyard telescopes will reveal a tiny pit crater on top of Omega Cauchy when seeing is excellent.

## MARE SERENITATIS

Located in the Northeast Quadrant of the Moon, **Mare Serenitatis** — the Sea of Serenity — contains a large number of interesting and unique formations. This mare fills a basin that was created by an asteroid impact some 3.87 billion years ago. Like all the other circular maria, the floor of Serenitatis is smooth and almost featureless, with only a handful of small craters. One of these is **Linné**, only a mile in diameter and located on the western side of the sea. It would be nearly invisible but for the white patch surrounding it. Before the space age, numerous observers recorded changes in the size and appearance of this object and as a result, Linné was once considered a good candidate for being an active lunar volcano. The controversy associated with this crater is discussed on page 48.

Another interesting feature on the floor of Mare Serenitatis is the famous **Serpentine Ridge**, running north-south for several hundred kilometers on the eastern side of the sea. The Serpentine Ridge is a wrinkle ridge, and it is one of the largest and most complex of the many that zigzag across the floors of all the mare lava fields. Sunrise on the Serpentine Ridge occurs some five days after New Moon and sunset happens four days after Full. West of the ridge is **Bessel**, the largest crater on the floor of Mare Serenitatis. A long single ray, possibly from Tycho, is associated with Bessel.

Barren though the floor of Mare Serenitatis is, the periphery contains a wealth of interesting formations. Imagine the mare as the face of a clock with due north representing noon. At 1 o'clock Mare Serenitatis spills into **Lacus Somniorum**, a smallish lava flow. And immediately to the north is the even smaller patch of mare called **Lacus Mortis** (photo on page 45). The 40-km crater **Bürg** is located just off-center within Lacus Mortis, and the remnants of Bürg's ejecta blanket are flung out in swaths heading to the northwest and the southwest. In between the swaths are two rilles which intersect to form a T.

Due east of Lacus Mortis is the crater-pair **Hercules** and **Atlas**, 70 km and 90 km in diameter respectively. Their rims and walls are similar, but their interiors could not be more different! The floor of Hercules has been flooded and has several interesting contrasts in albedo. A dark arc of lava can be seen on its northern end. A sharp-rimmed bowl-crater dominates the rest of the crater floor and may have obliterated any central peak.

The floor of Atlas is much rougher and lighter in color, although two dark spots are located there. The northern-most dark spot varies in shading significantly as the lunar month progresses. At times it is lighter than the southern spot, on other occasions they are nearly the same shade. A system of at least five delicate rilles crisscrosses the floor. These are best viewed with high magnification two or three days after Full Moon. The careful observer will also spot a miniscule system of central peaks within Atlas.

Moving back south to Mare Serenitatis, at 2 o'clock is the magnificent flooded crater **Posidonius** (photo on page 44) and its companion **Chacornac**, which butts up against its southeastern wall. Posidonius spans about 100 kilometers, while Chacornac measures half that. The floors of both craters are crisscrossed with numerous fine rilles. Posidonius has been flooded with lava to the point where it is almost filled to the rim. This can be seen on its western side, where the rim is almost but not quite breached. There is an interesting inner rim within Posidonius; it brightens considerably as the Sun rises high over the crater, giving it the appearance of a crater within a crater. In afternoon light, it can lend a spiral appearance to the crater.

At 3 o'clock is the 60-km remnant-crater **Le Monnier**, whose seaward wall has been destroyed by encroaching lavas. The second Russian automatic mobile laboratory, Lunokhod 2, roved far and wide across the dark floor of this crater. Not too far away, at 4 o'clock, you will find the stubby hills of the **Montes Taurus**, site of the Apollo 17 mission. This whole area is dark in appearance, making it easy to spot at Full Moon. Nearby, a series of arcuate rilles formed as a result of tectonic stresses caused by the filling with lava of the Serenitatis basin.

At 5 o'clock a gap in the ring that encircles Mare

**North of Aristoteles lies the eastern end of Mare Frigoris. On its northern shore, the mare's lava has invaded a complex terrain, flattening the landforms and destroying the interior terraces and central peaks of craters. Jean Dragesco photo.**

Serenitatis provides a connection to Mare Tranquillitatis. Sweeping in a long arc from 6 to 9 o'clock are the **Montes Haemus**, severely truncated hills rising no higher than 2,000 meters that were scoured by flying ejecta from the gigantic impact that formed the Imbrium basin to the northwest. The radial pattern created by the ejecta is quite apparent. On the southern edge of Mare Serenitatis, just off the tip of Haemus mountains, is the 40-km crater **Plinius**, which has a sharp rim. Immediately north of Plinius are three parallel rilles, an extension of the system in the Taurus mountains. In this locale is an interesting band of dark mare lava, which can be seen to encircle the entire perimeter of the mare. This dark lava represents older, iron-rich flows which in the center of the Mare Serenitatis are overlain with younger, lighter-colored titanium-rich lavas.

Finally, continuing the sweeping arc from 10 to 12 o'clock are the **Montes Caucasus**, which have also been scoured by Imbium ejecta. Through a small gap in the Caucasus mountains at 9 o'clock Mare Serenitatis connects with Mare Imbrium. When the terminator is passing across this gap, you can see a series of short wrinkle ridges looking like frozen waves in a sea of lava.

### Linné

The difficulty of interpreting visual (and sometimes even photographic) observations of the Moon is well illustrated by the case of the crater **Linné** (photo on page 43). This very small crater — only 2.5 kilometers in diameter — stands in the western expanses of Mare Serenitatis. It would be totally undistinguished if it weren't for a bright nimbus surrounding it, which makes it easy to spot on the dark expanse of the mare. When fully illuminated (from local mid-morning to mid-afternoon) the nimbus increases Linné's apparent diameter to almost 10 kilometers. At sunrise or sunset, however, Linné is merely a tiny circular object requiring high magnification to spot. But as the Sun's altitude increases it illuminates the nimbus until it grows so dazzling bright that the craterlet itself is totally concealed in its glare. You really have to witness this remarkable transformation to believe it.

Careful regular observations will soon show you that month-to-month changes in the angle of illumination cause significant alterations in the apparent intensity and shape of this nimbus. Once in a while this change can be dramatically different from the view presented just four weeks earlier. These remarkable variations in appearance led more than one 19th century astronomer to proclaim that Linné had undergone geologic changes — thus proving the Moon was still geologically active. This suspicion persisted well into the 20th century and even researchers who correctly dismissed volcanism as the dominant crater-forming mechanism were willing to concede that here there was a serious possibility of residual small-scale lunar volcanic venting. The hypothesis was that gases periodically escaping from Linné's old volcanic vent were laying down new whitish deposits around the crater.

These persistent reports prompted scientists to target Linné for close-up photography during the Lunar Orbiter missions. What these photographs revealed was nothing more than an extremely fresh impact crater with a perfectly formed bowl and an apron of light-colored ejecta. Not a shred of evidence for volcanism, past or present, was seen!

The impact that created Linné is apparently of very recent origin by lunar standards. This crater is probably no

A multitude of features await anyone viewing the waning gibbous Moon. Above: The Straight Wall looks almost artificial when you first see it. This fault scarp has its higher side to the east, so it appears as a dark line before Full Moon but becomes a bright line after Full. It lies in the center of an old crater nearly obliterated by lava flooding. Photos by Lick Observatory (right) and Donald Horne (above).

more than a few tens of millions of years old. Although we know that the whitish nimbus consists of pulverized rock or rock flour, we cannot explain why some small impacts result in a nimbus while others cause their ejecta to disperse for hundreds of miles creating ray systems. Possible contributing factors include the speed of the impacting projectile, the composition of the meteoroid, the angle of impact, and the composition and structure of the target terrain.

### MARE FRIGORIS and THE NORTH POLE

One glance at **Mare Frigoris** tells you this is not your typical lunar sea. Elongated in shape, the Sea of Cold is no doubt a compound feature consisting of two or three separate formations fused into one by lava flows of past ages.

The western end of Mare Frigoris is probably the remnant outer ring from the impact that created the Imbrium basin. The eastern portion, north of Mare Serenitatis, may be the coalesced remnants of two or more ancient impact basins. In total area Mare Frigoris about equals Mare Tranquillitatis. A number of interesting formations lie in the area of Mare Frigoris. The first of these, at the eastern end of the mare, is the 125-km crater **Endymion**, easily spotted because of its flooded dark floor. Endymion has thick walls and its floor is a virtually featureless expanse of frozen lava. This crater is useful in finding the next feature, **Mare Humboldtianum** (Humboldt's Sea), which straddles the lunar limb directly behind Endymion. The visibility of this formation, like that of Mare Australe to the south, is considerably influenced by lunar librations and may not be favorably placed for observation every month. About a week before Full Moon, the dark floor of Mare Humboltianum is easily seen. Its size varies greatly according to how much has been brought into view by librations. Immediately after Full Moon, the setting Sun casts long shadows in this area and you can observe, when librations are favorable, that two mountain rings encircle the formation.

Moving west across Mare Frigoris until we reach the point where it begins to become narrow, we encounter one of the most magnificent craters on the Moon, 90-km **Aristoteles** (photo on page 45). This is a fresh, almost perfectly preserved impact crater with a strongly textured ejecta apron full of radial hills and valleys surrounding it. (There is also a weak ray system associated with this crater.) The inner walls of Aristoteles are beautifully terraced and contain an enormous amount of detail. Two tiny central peaks can be seen on its floor, south of center. Tangent to Aristoteles on its eastern wall is the smaller crater **Mitchell**, and due south lies **Eudoxus**, located in the hills of Montes Caucasus. Slightly smaller than Aristoteles, 70-km Eudoxus forms an easily recognized pair with it. Eudoxus also has a interesting system of slumped terraces inside its rim.

North of Aristoteles, about midway between the northern shore of Mare Frigoris and the lunar pole, is the unique compound crater formed by 120-km **Meton** and two adjacent large craters. These have been flooded and give the appearance of a lunar 3-leaf clover. Directly west of Meton is one of the few sharp-rimmed craters in an ancient area otherwise dominated by greatly degraded formations. This is **Anaxagoras**, which comes into view a day or so after First Quarter Moon. Anaxagoras doesn't look much different from scores of similar fresh craters scattered across the lunar surface. But if you revisit this area at the time of Full Moon, you will see that Anaxagoras is the source of one of the major ray systems on the Moon.

Moving to the western end of Mare Frigoris, to a point marked by the sharp crater **Harpalus**, we come to the area where the mare spills into Sinus Roris. Here we have the opportunity to observe another fascinating crater, **Pythagoras**. Pythagoras is located about two-thirds of the way from Harpalus to the northwestern limb of the Moon, and becomes visible about a day and a half before Full Moon, as the terminator slides across it. This is a prominent crater similar in basic structure to Aristoteles or Copernicus, having terraced walls and a set of two matched

central peaks. What makes this crater particularly interesting is the view that is provided by its position close to the lunar limb. The impression one gets at the eyepiece is that of flying in a spaceship over the rim and looking obliquely into the bowl of Pythagoras. The effect is accentuated by the presence of the two central peaks. Pythagoras provides one of the few opportunities we have to see a well-preserved crater from a perspective other than the flat overhead view afforded by most other lunar craters.

## THE SOUTHERN HIGHLANDS

The **Southern Highlands** region is a very complex and difficult area to observe due to the enormous number of craters found there. It is an ancient region, preserving traces of some of the oldest terrain on the Moon. Extending in a wedge from the south polar region north to the equator in the center of the lunar disk, the highlands contain such well-known craters as Clavius and Tycho to the west, Janssen and Piccolomini to the east, and Ptolemaeus to the north. In between is a vista of unrelenting crater-upon-crater. Some of these have interesting features if one takes the time to locate and identify them.

A good starting point is **Thebit**, a 55-km crater located on the southeastern shore of Mare Nubium not far from the Straight Wall. A reasonably fresh crater, Thebit has a rim interrupted by the small crater Thebit A, giving it a diamond-ring effect. Thebit A is in turn interrupted by the craterlet Thebit L, providing a nice grouping. Immediately to the southeast is the large and much older crater **Purbach**, twice the size of Thebit. It encroaches on a similar sized crater, **Regiomontanus**, which has the shape of a squashed melon as a result. Regiomontanus is notable for its large central peak, located where the center would have been before Purbach deformed the crater. This peak has an unusually large craterlet superimposed on its summit making it look exactly like a volcano. In fact this is merely the result of a chance impact.

Due east of Regiomontanus is the 70-km crater **Werner** with its high, terraced walls. Werner is one of the freshest craters in this region. Two Werner-diameters directly to the north is the heart-shaped crater **Delauney**, which is divided in the middle by a ridge. Arcing to the northeast from Delauney is a set of five matched craters, each having well-preserved central peaks despite being of apparently different ages. This appealing grouping consists of (south to north) **Faye**, **Donati**, **Airy**, **Argelander**, and **Vogel**. Each crater is 30 to 35 kilometers in diameter and has much interesting detail.

Return now to Purbach. Approximately five Purbach-diameters due east is the easily spotted **Sacrobosco**, whose identification is made simple by the characteristic group of three deep bowl-craters located on its floor. Sacrobosco is the jump-off point for locating the crater pair **Abenezra** and **Azophi**, which lie just to the northwest. Abenezra and Azophi share a common wall/rim where they adjoin. Particularly noteworthy is the way Abenezra, the northernmost of the two, has thrown huge amounts of rubble onto the floor of an older crater to its southwest.

Moving five crater diameters due south of Sacrobosco, we come across a pair of large craters consisting of 110-km **Maurolycus** on the east and 135-km **Stöfler** on the west. Both are part of a complex grouping. Maurolycus is an old crater superimposed on an even older unnamed crater, of which only a small part of the wall is left, visible to the south. Stöfler is noted for its smooth floor, which is

**The Southern Highlands are a tangle of craters unmodified by anything except later impacts. Clavius (upper center) contains a rich array of craterlets within it. Christian Arsidi photo.**

crossed by numerous light streaks, most likely belonging to the ray system from Tycho. The eastern wall of Stöfler has been destroyed by the 70-km crater **Faraday**, which in turn has been severely battered by a series of smaller craters. One can trace at least four "impact generations" here by noting the juxtaposition of craters in this area. South of the Stöfler crater is **Heraclitus**, another much-modified crater with a central ridge dividing it.

## Clavius

At 225 kilometers in diameter, **Clavius** ranks as one of the largest craters on the near side of the Moon. The only larger craters are the almost completely destroyed **Deslandres**, located due north, and the equally ancient crater **Bailly**, located just to the southeast of Clavius on the lunar limb. Deslandres is so poorly preserved that few observers even recognized it as a crater until the 20th century, and Bailly can be observed well only when the librations are favorable. In contrast, Clavius — while still old as craters go — is well preserved and located for easy observation each month. It is notable for its thick, rugged walls, which rise high above a relatively smooth convex floor. Located on the floor is a chance alignment of five craters that forms a sweeping arc with each successive crater proportionally smaller than the last.

The largest of these, 50-km **Rutherfurd**, nestles just inside the southeast edge of Clavius' rim. A fresh crater, Rutherfurd has a system of unusually large central peaks, and its ejecta blanket has created an interesting tracery to the north on the smooth floor of Clavius. Making a

matched pair with Rutherfurd is the similarly-sized **Porter**, which stands on Clavius' northeast wall. (Older books give its name as Clavius B.)

Following the arc along from Rutherfurd, we come in turn to the bowl-craters **Clavius D, C, N,** and lastly, **J.** Just to the southwest of Clavius C is a miniscule group of central peaks — all that Clavius could muster, apparently. It is interesting to note that an astronaut standing near these peaks would be totally unaware of the huge ramparts that form Clavius' outer wall. The curvature of the Moon is so great that these walls, being over 100 kilometers distant, would lie well below the apparent horizon. When steady seeing prevails, up to two dozen craterlets can be seen littered across the floor of Clavius.

Because of its size, Clavius is the only crater that can be detected by the naked eye, but you'll need special circumstances to do it. If you look at the Moon just one day after First Quarter, the terminator should be passing across Clavius. At local sunrise, the high walls prevent the grazing sunlight from illuminating the floor of Clavius, and so the interior remains inky-black for a while after sunrise. At the same time, the rim and surrounding high-

Cracked with rilles, the region of Mare Vaporum lying next to the crater Triesnecker is fascinating to explore at high magnification on a night with steady seeing. Above: Mare Vaporum is a relatively small expanse of lava tucked between Mare Serenitatis and Mare Imbrium **(south at right). Photos by Jean Dragesco (right) and Lick Observatory (above.)**

land areas are brightly lit. Precisely at these moments, Clavius will appear as a dark notch in the bright terminator. When this notch reaches its maximum size, just before first sunlight streams onto the floor of Clavius, it can be readily detected by a prepared observer with no optical aid.

One other interesting and beautiful effect is associated with sunrise on Clavius and can be seen telescopically at the same time you make your naked-eye observations. As the Sun rises higher over Clavius, the rims of Rutherfurd and Porter become illuminated first and present the appearance of bright rings suspended in an inky black sea. In turn, the same effect can be seen with each of the small craters forming the arc across the floor of Clavius.

Clavius is well-preserved, but clues point to its real age, which is great. Evidence of ejecta from the impact, which should be visible in the surrounding highlands, has been largely obliterated, and only traces of terracing remain on the inner walls of Clavius. Its floor, by virtue of its smoothness and the small size of the central peaks, suggests that resurfacing has occurred since the impact that created it.

## Tycho

Of all the large craters on the Moon, 85-km **Tycho** is one of the most recent (photo on page 49). No matter when during a lunation you observe it, Tycho reveals the distinctive characteristics of an extremely fresh crater: rays and sharp details. Tycho was extensively studied by the Lunar Orbiters and Surveyor 7, and it is also fairly certain that samples of its ejecta were obtained at the Apollo 17 landing site thousands of kilometers away. These date the Tycho impact to 109 million years — very recent in lunar geological terms.

Situated in crowded terrain to the north of Clavius, Tycho would be difficult to spot at sunrise were it not for the magnificent system of lunar rays emanating from it. At sunrise, when long shadows make even minor features stand out boldly in relief, it is hard to distinguish Tycho from neighboring formations. But the long rays extending to the east point back to their source and you need only follow these signposts.

The first obvious thing you'll notice is that the area immediately surrounding Tycho seems lower. (Some observers have suspected that Tycho is sitting in the much-altered remains of an ancient crater about the size of Clavius.) Anyway, as the local day brightens toward noontime at Full Moon, this area takes on the appearance of a dark ring enveloping the brighter Tycho. This feature is probably dark material from deep within the Moon that was excavated by the Tycho impact. The rest of the ejecta blanket extending even further out from the rim has severely disrupted the surrounding terrain. Most striking is the enormous system of secondary craters in the area surrounding Tycho, visible at high magnification. When the Sun rises a little higher, Tycho's massive central peak comes into view, first appearing as a starlike point of light in the jet-black crater bowl. Later in the lunar day the Sun reveals an extensive terrace system dominating the inner walls.

Tycho's rays are the Moon's most prominent and well worth exploring. On any given night the relative intensity of these rays will vary, but eight or so generally stand out. Trending northwest is the famous "double-ray," which extends into Mare Nubium to the vicinity of the crater Bullialdus. Moving in a counterclockwise direction from the double-ray, we note a large zone with only short, weak rays. This occupies an arc of almost 120°. We then come to a bright ray extending southwest toward the crater Scheiner. Close by is another bright ray that crosses Clavius and then heads toward the south polar region. Continuing in a counterclockwise direction, we come upon another long ray which can be traced eastward all the way to the crater Fracastorius, located on the southern shore of Mare Nectaris in the eastern hemisphere. Lastly, at least two more bright rays trend northeast toward Mare Tranquillitatis and Serenitatis. (The Serenitatis ray crosses the small crater Bessel.) Near the middle of Deslandres a bright patch known informally as Cassini's Bright Spot lies in the path of one ray.

Interdispersed with these eight major rays is a welter of shorter ones. As the angle of illumination changes from night to night, these take turns at becoming relatively brighter or dimmer. Under excellent observing conditions it is possible to begin resolving portions of some of these rays into tiny craterlets with sprays of ejecta, which represent secondary impacts from the Tycho event. But in general the rays are as wispy and amorphous as a cloud in the sky. They defy every attempt to resolve detail and all but disappear from view under high magnification.

## MARE VAPORUM

**Mare Vaporum**, the Sea of Vapors, is located close to the center of the lunar disk. Hemmed in on all sides by encroaching younger formations, its circular shape is nonetheless apparent, especially when viewed near Full Moon. But virtually all traces of Mare Vaporum's rim have been obliterated by Mare Serenitatis to the northeast, Mare Tranquillitatis to the east, Sinus Medii to the south, Sinus Aestuum to the west, and the Apennine mountains to the north.

The **Montes Apenninus** form the southeastern portion of Mare Imbrium's basin rim. The Apennine slopes that gradually lead down to the smooth lava plains of Mare Vaporum are a study in twisted, tortured terrain. On the other side, facing Palus Putredinis and Mare Imbrium, the mountains plunge precipitously to the lava floor below. A succession of high peaks are strung along its length. To the east of Mare Vaporum are the much older Montes Haemus, the southwest portion of Mare Serenitatis' basin rim. In contrast to the lofty Apennines, these consist of severely truncated, stubby hills. A pattern of valleys pointing toward the Apennines and Mare Imbrium beyond demonstrate that Montes Haemus were scoured by ejecta from the Imbrian impact. South of Montes Haemus is the most outstanding crater on Mare Vaporum, **Manilius**. This young, 40-km crater has an interesting group of central peaks and a small ray pattern, and it is quite bright when illuminated at Full Moon.

Due south of Manilius are two magnificent rilles, **Rima Ariadaeus** and **Rima Hyginus**, which are described on page 56. The highland area between Manilius and the Ariadaeus rille contains a series of elongated valleys with smooth dark floors. The formation of these is also linked to the Imbrian impact, as indicated by their alignment. The largest of the valley-like formations includes the battered remains of a dark-floored, ancient crater 90 kilometers across named **Julius Caesar**. Nearby is **Boscovich**, which has been similarly affected and has an elongated

**A curious trio of craters lies along the shore of Mare Nubium. The flat floor of Ptolemaeus (bottom) has many details visible only under grazing sunlight. Gerard Therin photo.**

valley to the north. The floor of Boscovich is crossed by two fine rilles.

A narrow band of lava connects Mare Vaporum with **Sinus Medii** to the south. On the smooth surface of Sinus Medii, just east of the 25-km crater **Triesnecker**, is one of the most complex systems of branching and overlapping rilles to be found on the Moon (photo on page 53). Much more delicate than the nearby Rima Hyginus, the Triesnecker rilles show a dozen or more spider-webs which will pop into view during periods of steady seeing. Triesnecker is also the center of a minor ray system.

Moving westward from Triesnecker across Sinus Medii, we encounter two 30-km horseshoe-shaped craters, **Schröter** and **Sömmering**. In both cases their rims have been breached by lava on the south side. Due west of Schröter, on the plains of Sinus Aestuum, we come across two perfectly shaped bowl-craters with razor-sharp rims. These are designated **Gambart B** and **C**, after the 25-km flooded crater located further to the west. Each bowl-crater measures 12 kilometers across. Between them stands a large flat dome with a curious appendage on its southern end.

### Rima Ariadaeus and Rima Hyginus

It's fortuitous that two of the most attractive rilles on the Moon are located close to the center of the lunar disk where they can be observed in detail. Rima Ariadaeus and Rima Hyginus are both generally oriented east-west and taken together they span almost 1/12th the distance from limb to limb. Sunrise on the easternmost portion of **Rima Ariadaeus** occurs about six days after New Moon and it takes a full day before the terminator moves far enough to reveal the westernmost point of Rima Hyginus.

The eastern end of Rima Ariadaeus lies on the border of Mare Tranquillitatis, west of the crater Arago. The rille proceeds across a highland region and ends at the border of Sinus Medii. Curiously, both ends of the Ariadaeus rille are forked. The rille is a fine example of a graben, the geological name for the dropped terrain lying between two roughly parallel faults. Spend time tracing the rille; you'll see it cutting across hills and ancient crater rims unimpeded. At one spot it is offset by its own width, which appears to be the clearest example of a strike-slip fault found on the Moon. The Ariadaeus rille is probably also the only lunar rille in which you can view details. These appear on the floor of the rille, but require good seeing, moderate aperture, and high magnification.

**Rima Hyginus** presents a even more interesting appearance and some unexplainable details as well. Structurally, it must relate to Rima Ariadaeus because at its eastern extremity they run parallel and are only 35 kilometers apart. But there are interesting differences. About halfway along its length, the Hyginus rille intersects the small, rimless crater **Hyginus** which is 11 kilometers in diameter. There the rille turns 30° to the north and proceeds straight again from that point. There is no apparent explanation for why Rima Hyginus should have diverted from its straight path in this way. The most striking difference between the two rilles is that Rima Hyginus contains many craterlets on its floor, strung along its length like pearls on a string. These are most obvious in the vicinity of the crater Hyginus. The craterlets are also rimless and are probably places of localized subsidence along the fault line that created the rille itself.

Virtually all lunar rilles demand grazing light to be seen — it's needed to throw their features into relief. But

Rimae Ariadaeus and Hyginus can also be spotted under high lighting, even at Full Moon, when they appear as thin white streaks on the lunar surface. Knowing what to look for and their general location aids in detecting them.

## MARE NUBIUM

**Mare Nubium** — the Sea of Clouds — is a circular sea located in the southwest quadrant of the Moon. It might aptly be called the "Sea of Ghost Craters," as many of these abound on its floor, some becoming apparent only under the high lighting of Full Moon. In addition, Mare Nubium has a good selection of mostly destroyed craters — these include **Lassell**, **Wolf**, **Guericke**, **Parry**, **Gould**, **Opelt**, **Kies**, and **Lubiniezky**. This suggests that the Nubium basin is very ancient and that it was partially covered with ejecta from other basin-forming impacts before the era of volcanism began and filled it with lava. In most of the lunar seas, the central areas are featureless because the basins are deepest there and the lava has obliterated all trace of the basin's bottom topography. In the case of Mare Nubium, however, the central region must not have been too deep, because when the lavas began flowing they did not completely engulf craters such as Gould, thus leaving the highest ramparts of its rim intact.

The rim of Mare Nubium has been destroyed on its northern side. Adjacent to the eastern shore of Mare Nubium is a chain of three craters located almost exactly on the Moon's central meridian. These come into view just at First Quarter Moon. From north to south, we find **Ptolemaeus**, a very ancient flooded crater, **Alphonsus**, a moderately old crater with remnants of a central peak, and **Arzachel**, a relatively fresh and well-preserved crater. This famed group — rich in features to explore and often observed — is described in detail beginning on page 59. Due west of Alphonsus, on the floor of Mare Nubium itself, is the 35-km crater **Davy** and its giant companion, the flooded ancient remnant **Davy Y**, more than twice Davy's size. On the otherwise featureless floor of Davy Y is an interesting straight line of craterlets. Are these a chain of volcanic vents, such as is commonly seen on Earth, or are they a string of secondary impacts from a nearby (but so far unidentified) crater? The actual origin of this crater chain is still unresolved.

Straight south from Davy is the lunar showpiece, **Rupes Recta**, or the Straight Wall. Approximately 130 kilometers long, this stunning and unique formation is described on page 60. On the southern shore of Mare Nubium is the large flooded crater **Pitatus**, some 100 kilometers in diameter. The remnants of a central peak are visible, and the careful observer will be able to detect a series of rilles that run parallel to the inner rim of Pitatus on its northern side, where the wall is actually breached. Butted up against the western wall of Pitatus is **Hesiodus**, focal point for two remarkable formations. The craterlet **Hesiodus A** sits astride the southwest rim of Hesiodus and is the finest example of a crater with double concentric walls on the Moon. Seen under the grazing light of sunrise or sunset, this delightful little feature looks just like a lunar bullseye — one crater rim nestles perfectly within the other. The most likely explanation of this rare type of formation is that by chance a small impact landed directly in the center of a preexisting bowl-crater. Near the north-

**Splashed and streaked with rays from fresh impact craters, the face of the Last Quarter Moon is largely covered with lava flows. Mt. Wilson Observatory photo.**

58

Its smooth dark floor looking like a lake, Plato stands amidst the lunar Alps mountain range. Eastward from it the Alpine Valley, a geological graben, chops through the mountains. Above: Palus Putredenis, where Apollo 15 landed, lies in Mare Imbrium between the Apennine mountains and Archimedes crater. Photos by Jean Dregesco (left) and Lick Observatory (above).

western corner of Hesiodus begins the wide straight rille **Rima Hesiodus**, which can be traced in a southwestern direction across the smooth lavas of Mare Nubium into rough highland terrain and back out again onto the lava flows of Palus Epidemiarum.

Moving to the west of Hesiodus, we next encounter the twin craters **Mercator** and **Campanus**, both about 50 kilometers across and placed on the rim of Mare Nubium. On the mare floor to their northeast is the crater remnant **Kies**. All that remains of it is the topmost part of Kies' rim, which makes a beautiful broken ring as the Sun rises or sets over it. Directly west of Kies is the isolated dome **Kies Pi**, which has a clearly visible pit crater on its summit. Kies Pi is one of the most perfectly formed domes you can see on the Moon. Due north of Kies is the impact crater **Bullialdus**, its splash-pattern of ejecta clearly displayed on the floor of Mare Nubium. This crater is a scaled-down Copernicus and is every bit as interesting to examine for all the fine details — terraces, central peak, rough blanket of ejecta — that it possesses on its floor and rim.

The northwest portion of Mare Nubium is bounded by the **Montes Riphaeus**, the pathetic remains of what must have once been magnificent basin walls. These now look much like a lobster's claw and their peaks barely exceed a kilometer in altitude. On the western side of Riphaeus mountains is the small bowl-crater **Euclides**, only 13 kilometers across but easily identified by the bright nimbus surrounding it. This bears watching as the Sun rises higher over it. Brilliantly white even just after sunrise, by the time of Full Moon Euclides evolves into one of the brightest spots on the Moon. Once you know its location, Euclides is one of the few craters instantly identifiable in the glare of Full Moon.

The portion of Mare Nubium lying to the east of Montes Riphaeus was renamed **Mare Cognitum**, the Known Sea, when the unmanned Ranger 7 probe crash-landed here in July 1964. Before impact, the probe took the first closeup photos of the Moon's surface. Mare Cognitum is seemingly uninteresting as viewed through amateur telescopes, but this area was nonetheless of great geological interest to the scientists who carried out the planetary exploration program of the 1960s and 1970s. After Ranger 7, Mare Cognitum was visited by Luna 5 a year later. In 1967 Surveyor 3 made a successful soft landing there and 2½ years later, the Apollo 12 crew set down their Lunar Module next to Surveyor 3. Walking over to the now-inert Surveyor probe, the astronauts retrieved various pieces of the lander to bring back to Earth. (These "souvenirs" told scientists what happens to hardware when exposed to lunar conditions for many months.) Finally, in 1971 Apollo 14 landed to the east just outside the rim of the crater Fra Mauro. This region remains the most intensely explored area on the Moon.

Despite its severe degradation, the Fra Mauro crater complex is worthy of telescopic exploration. Three ancient craters abut there, seemingly joining in a single compound formation. In addition to **Fra Mauro**, 95 kilometers wide, there are **Bonpland** and **Parry** to the south, respectively 60 kilometers and 45 kilometers across. Because these craters are so eroded you have to view them when the terminator passes across them. Once the Sun gets high, the relief is too low to be easily seen. Even a day after local sunrise is enough to obliterate most detail. But notice how the three crater walls merge in the center of the formation. Can you see from the way they overlap that Fra Mauro is the oldest and Parry the youngest of the three? A rille system cuts southward across all three craters. Starting in the north, a rille bisects the floor of Fra Mauro. As it trends south, the rille forks about two-thirds of the way across Fra Mauro's floor. The western fork heads into Bonpland, where it bisects that crater. The eastern fork goes into Parry and runs close to its western rim. Notice how the rilles make notches in the rims as they pass from Fra Mauro to the other two craters to the south.

## Ptolemaeus, Alphonsus, and Arzachel

This trio of craters, each a different size and age, is perfectly placed for observation at First Quarter Moon. Seen at sunrise they form a distinctive group as they emerge from the long lunar night. The northern member of the trio is **Ptolemaeus**, the largest (150 kilometers across) and oldest (photo on page 55). Its rim, whose walls are only a little over a kilometer high, is heavily degraded and peppered with subsequent small impacts. Its floor is smooth, save for a handful of small bowl-craters, having been resurfaced with plains deposits generated by the Imbrium impact. To the east of Ptolemaeus is **Albategnius**, which is older and consequently more battered. Then, north of Albategnius is the ancient crater **Hipparchus**, whose rim is all but destroyed on its western side. The area around these craters has many long thin valleys which are gashes caused by ejecta from the Imbrium impact.

Ptolemaeus offers an extremely interesting observing opportunity. If you catch this formation when the terminator has moved far enough west to illuminate its eastern rim but the Sun is not yet high enough to illuminate its floor, you will be treated to watching a sunrise on this crater. Initially you'll see the first rays of sunlight on the western end of Ptolemaeus' floor. Then the jagged shadows cast by peaks on the eastern rim will lie in streaks across the 150 kilometer expanse of the floor. During the next hour and a half the shadows will rapidly shorten until the whole of the floor is sunlit. The speed with which this progresses tells you how little elevation the rim actually has. As the shadows march across the floor of Ptolemaeus, several extremely shallow saucer-shaped craters will come into view. These rimless craters have so little relief that they can be seen only at sunrise and sunset.

Butted up against the southern end of Ptolemaeus is the smaller and younger crater **Alphonsus**, whose northern end overlays and has destroyed a piece of the rim of Ptolemaeus. Alphonsus is 120 kilometers in diameter. Its walls are higher and better preserved than those of its northern neighbor, and it still has a remnant central peak. Two fine rilles can be detected on the eastern half of Alphonsus' floor. But the most distinctive feature of Alphonsus is best seen after the Sun rises high over this formation. It consists of three dark spots arranged in a large triangle on the floor of the crater. These spots have long been suspected of being volcanic in origin and although Ranger 9 was sent into Alphonsus to explore this possibility — and despite much orbital photography — the issue remains open.

Alphonsus has also been the site of suspected lunar transient phenomena (often called LTPs). In general, these are anomalous glows or hazes reported from time to time. In the case of Alphonsus, its floor has been said to be occasionally obscured and in 1957 a Soviet astronomer reported gases venting from Alphonsus' central peak. So far, however, no convincing physical evidence for LTPs has yet emerged and the subject is uncertain enough that many planetary geologists flatly deny LTPs exist.

South of Alphonsus is the third member of the meridional trio, **Arzachel**. Nearly 100 kilometers from rim to rim, Arzachel is fresh. It has a sharp, high rim, a highly developed system of terraces along its inner wall, and a large central peak. A rille, wider than the one in Alphonsus, can be seen on the eastern half of the floor, close by a bowl-shaped crater. When you compare how long it takes the rising Sun to illuminate fully Arzachel's floor compared to Alphonsus and Ptolemaeus, you'll see that Arzachel's rim is substantially higher. Nestled between Arzachel and Alphonsus on their western side is the small crater **Alpetragius**, only 40 kilometers wide. On its floor is a huge central peak, some 20 kilometers across, whose disproportionate size gives this formation the appearance of an egg in a nest.

### The Straight Wall

One day after First Quarter Moon, the Sun rises on a remarkable feature. **Rupes Recta** — the Straight Wall —lies near the eastern edge of Mare Nubium (photo on page 48). The grazing light of sunrise reveals that it nearly bisects a very ancient ghost crater, one of many found on the surface of Mare Nubium. At sunrise, the Straight Wall throws a wide black shadow onto the smooth lava to the west. This tells you that the eastern side of the formation is higher. At first the shadow is so wide that it can be spotted in the smallest telescope. But as the Sun rises higher, the shadow narrows until it all but disappears from view a few days before Full Moon. But the Straight Wall reappears a few days before Last Quarter Moon as a thin bright line which is harder to spot than the shadow. The bright line is the result of the Sun's rays striking the face of the wall more directly as the Sun moves lower in the Moon's west.

Careful examination of the dark shadow seen after local sunrise, using medium magnification, shows that the wall is not nearly as straight as its name might imply. Nor, we

have further discovered, is it really even a wall. Analysis of the geometry of the Straight Wall's shadow demonstrates that it has a maximum elevation of only 600 to 800 feet above the surrounding terrain. And what is even more surprising, the "wall" itself slopes at only about 40°. The name fools you. Early lunar astronomers were misled by their instruments into picturing something like a sheer cliff. Don't blame them too much — when you observe the Straight Wall through your telescope, you too will find it hard to believe that it isn't a sheer cliff!

Even stranger is the fact that if you were standing on the Moon a few kilometers west of the base of the Straight Wall, it's entirely probable that this formation would be difficult, if not impossible, to detect. This is a phenomenon that the Apollo astronauts had to contend with on several occasions. Formations that were obvious as seen from overhead under grazing illumination were sometimes impossible to find on the ground. One of the best examples was the time the Apollo 14 explorers were unable to locate Cone Crater, the major objective of one particular moonwalk. Later analysis of their track showed that Alan Shepard and Ed Mitchell had actually been standing right on the outer slope of Cone Crater but were unable to recognize it as such.

Immediately to the west of the Straight Wall and still within the circle of the large ghost crater, is the fresh bowl-shaped crater **Birt**, only 17 kilometers in diameter. Butted against its wall is the smaller bowl-crater, **Birt A**, which is only 7 kilometers across. Moving a small distance to the west of Birt, we come across a fine, slightly winding rille running parallel to the Straight Wall. Perhaps the two are geologically related. Larger scopes will reveal the presence of small craterlets at either end of this rille. The northernmost craterlet, named **Birt E**, was probably the vent for the lavas that created the rille. This craterlet is only 3 kilometers wide. The craterlet on the southern tip of the rille is even smaller (less than 2 kilometers) and is a good test of resolution.

## MARE IMBRIUM

The Sea of Rains, **Mare Imbrium**, is for many observers simply the most magnificent of all the lunar seas. Since the basin that contains the sea is one of the more recently formed (albeit the events took place 3.85 billion years ago), its features are relatively well preserved. For example, you can find ample evidence of the ejecta thrown out of the Imbrium basin during the great impact of the 100-kilometer asteroid that created it. Radial valleys gouged by the ejecta abound in the Montes Haemus and Mare Vaporum areas to the southeast and in the region surrounding the crater Ptolemaeus further to the south. Use a low-power eyepiece around First Quarter and you can trace the gouges in the lunar crust out from their source in Mare Imbrium over much of the visible Moon.

Ample indications suggest that in its original condition, the Imbrium basin took the form of a three-ring bullseye. Today we say that Mare Imbrium is bounded on the north by the **Montes Alpes**, to the southeast by the **Apennines**, and to the south by the **Montes Carpatus**. In fact, this broken circle is really just the middle ring of the original three. The innermost ring has been almost entirely inundated by the lavas emplaced during the hundreds of millions of years that followed the basin-making impact. The few traces of this inner ring that remain consist of isolated bits and pieces of the highest peaks of the original rim. Starting in the north and moving clockwise, we find **Montes**

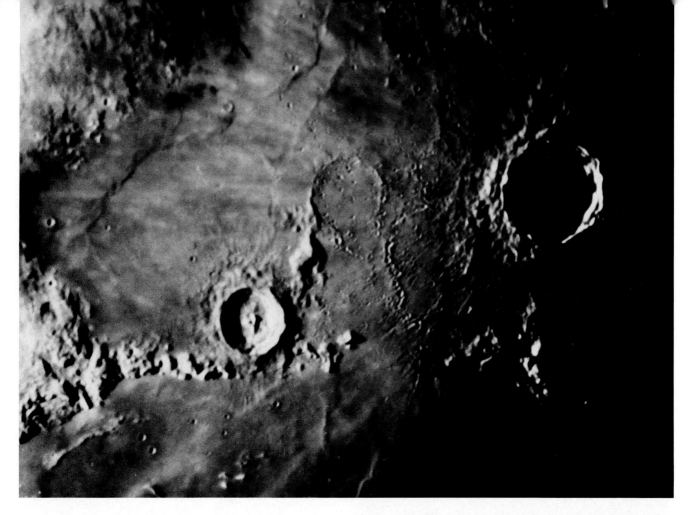

Recti, **Montes Teneriffe**, the isolated single peaks **Pico** and **Piton, Montes Spitzbergensis** (at 4 o'clock), the massif **La Hire** (at 7 o'clock), and to the west, a large wrinkle ridge arcing northwards from the craters **Diophantes** and **Delisle** through **Caroline Herschel** and up to **Sinus Iridum**. The outer ring of Mare Imbrium is defined by Mare Frigoris to the north and very likely Sinus Aestuum to the south. But most of the evidence for this ring is tenuous, having been obliterated by the encroaching lavas of the peripheral mare areas.

Mare Imbrium is the focal point for many interesting and unique lunar features. Once again let's picture the mare as the face of a clock with 12 noon to the north and quickly survey the best of these. Exactly at noon is the dark, flooded crater **Plato**, a frozen lake of lava discussed in detail on page 63. Moving east through the lunar alps we next come across the **Alpine Valley**. This lunar showpiece is also described separately (page 64). Continuing along the Montes Alpes we encounter **Mons Blanc**, rising 3,500 meters above Mare Imbrium. Then, shortly thereafter the mountain chain abruptly ends at the tall **Promentorium Agassiz**. There we encounter the ruined and flooded crater **Cassini**, whose characteristic appearance is caused by its floor being punctured with two bowl-craters. Unlike some other flooded craters (for example, Reiner in Oceanus Procellarum) the exterior ejecta of Cassini is still well preserved and contains much detail. Due west of Cassini is the isolated peak **Piton**, 2,000 meters high, whose long shadow is interesting to follow at sunrise.

To the south of Cassini and Piton are the crater pair **Aristillus** and **Autolycus**. Aristillus, the northernmost, has a magnificent system of splashed ejecta, which shows

**Perhaps the Moon's most dramatic sight is sunrise over Copernicus, an event which takes several hours to occur. Low Sun also reveals the ruined buried form of Stadius lying between Copernicus and Eratosthenes and a network of secondary impacts from Copernicus. Jean Dregesco photo.**

particularly well against the smooth lava of Mare Imbrium. The crater, 55 kilometers across, has unusually thick walls and a well-formed central peak complex. Aristillus is also the center of a weak but clearly defined lunar ray pattern. These rays have faded almost to the point of invisibility. Tangent to Aristillus at its northern end is a ghost crater, best seen as the lunar terminator passes across it. Autolycus, about 60 percent the size of its partner, provides an interesting contrast. It has a poorly defined ejecta system, a thin rim, and no central peaks.

Immediately south of Autolycus, occupying 4 o'clock on our imaginary timepiece, is **Palus Putredinis** (photo on page 59) and its neighboring large crater Archimedes. This is a significant geological area, perhaps preserving the only piece of original Imbrium basin not totally flooded with lava. Perfectly placed near the center of the lunar disk, it contains numerous points of interest, which are discussed beginning on page 64.

Moving west from **Archimedes** — a flooded 80-km showpiece with a flat, almost featureless floor — toward the interior of Mare Imbrium, we arrive at the first of four noteworthy craters. **Timocharis** is a perfectly formed medium-sized (35-km) crater with a sharp rim, terraces, and a well-defined ejecta pattern. Its central peak has been largely obliterated by a craterlet. Due west lies the slightly smaller crater **Lambert**, which is the focal point for a large intersecting system of wrinkle ridges. Lambert has prominent terraces and a crater superimposed on the spot

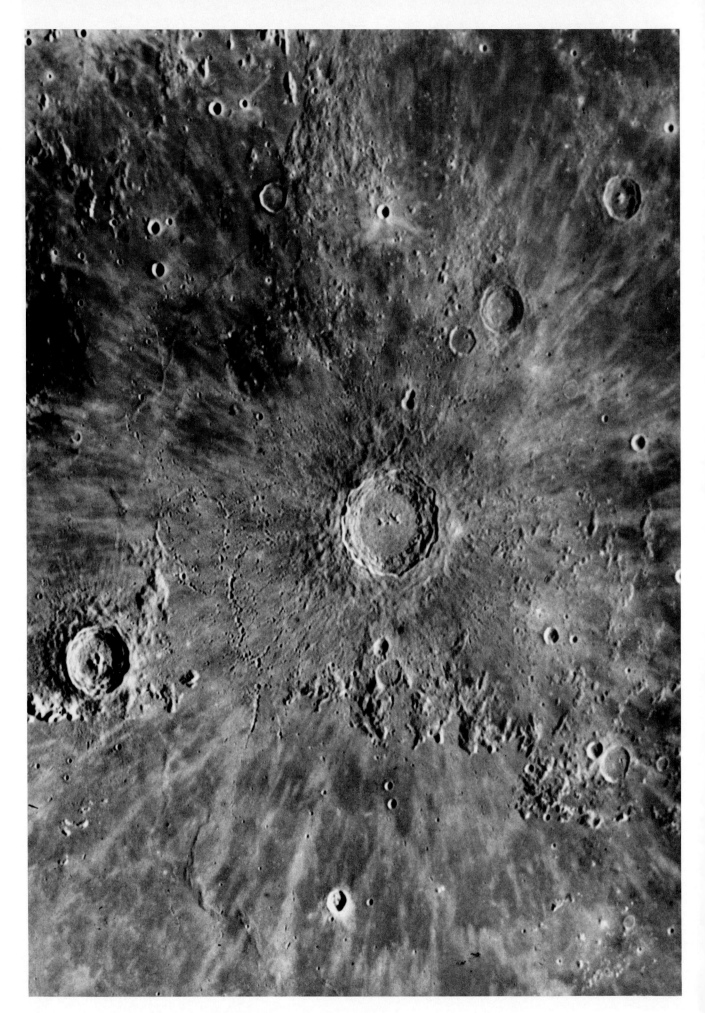

where the central peak once was. Immediately south of this crater is the ghost crater **Lambert R**, visible only when the terminator is close by. Further south is the still smaller crater **Pytheas**, which is extremely bright and easily found at Full Moon, when most craters are washed out and indistinct. Somewhat further west is the crater **Euler**, which has more structural detail. To the north of Euler and Lambert is the isolated massif **La Hire**, rising 1,500 meters above the mare floor.

Returning to the perimeter of Mare Imbrium at 5 o'clock, next to the southwest tip of the Apennines, is the well-formed crater **Eratosthenes**, which all but disappears from view at Full Moon, unlike most relatively fresh craters. A bit further to the west is the buried formation named **Stadius**, discussed on page 67. At 6 o'clock, just south of the Carpathian Mountains, is the most magnificent crater on the Moon. Perfectly preserved, having abundant detail of all kinds, and ideally placed for telescopic examination, **Copernicus** is truly the gem of the Moon. It was formed when a high-speed projectile perhaps only 2 kilometers wide struck the Moon with a force estimated to be equivalent to a million 1-megaton H-bombs. A detailed guide to Copernicus begins on page 67.

Finally, there is a large gap from 7 o'clock until 11

o'clock, where Mare Imbrium and Oceanus Procellarum spill into each other. Then we arrive at the grandest of all lunar bays, **Sinus Iridum** (see page 68) completing our circle of this large lunar sea. Don't neglect to examine Mare Imbrium at Full Moon and to note the lighter and darker areas across its smooth expanse of lava. These do not necessarily correlate to formations having relief, such as craters or wrinkle ridges, and largely show changes in the lavas that flowed in several stages over many millions of years to fill the Imbrium basin.

## Plato

The flooded crater **Plato** is located in the northwest quadrant of the Moon in a narrow strip of highland area between Mare Imbrium and Mare Frigoris (photo on page 58). Some 100 kilometers in diameter, Plato has a dark floor, making its location obvious. At low power, the floor looks as smooth as chocolate pudding and completely featureless. However, under excellent seeing conditions you might be able to spot three or four tiny craterlets at high power. In fact, detection of these craterlets is a good test of just how steady seeing really is.

Like Alphonsus, Plato is an area of reported lunar transient phenomena. Some observers have noted that one or more of these craterlets are obscured temporarily while at the same time others are easily detected. According to the hypothesis, local venting of gases may create dust clouds that cause these brief obscurations. While there's no final word yet on LTPs, it would be wise to remember that such phenomena are seen only rarely — and since the effects always seem to hover right at the limit of detectability, some skepticism seems in order. On the other hand, the steady march of the jagged shadow cast by the eastern wall of Plato on its smooth floor, seen as the Sun rises on

**The far-flung rays of Copernicus cross mountains and craters alike. They become most visible when the Sun stands high over the region — in local terms, from midmorning until midafternoon. Above: West of Plato on Mare Imbrium's northern edge, Sinus Iridum is a large crater with a breached wall on the mare side. It formed after the Imbrium basin impact but before the lava floods had filled the basin. Photos by Lick Observatory (left) and Jean Dragesco (above).**

this formation, is something you can witness every month. Look for this on the eighth day after New Moon. As with Ptolemaeus, you may be surprised at how rapidly the shadow moves as the Sun rises. Even 15 minutes are enough to change the location and shape of the shadow noticeably.

When you tire of Plato you'll find your telescope is already pointed at a region of great diversity. Move west along the narrow strip of highland and you will soon arrive at **Sinus Iridum**, the Bay of Rainbows. Move in the opposite direction and the famous **Alpine Valley** will pop into view. Both formations are described separately below. The vast expanses of Mare Imbrium, directly south of Plato, contain several unique features. Just to the southwest are two mountain formations poking through the lavas of the mare. Closest to Plato are the **Teneriffe** mountains, and a little further west is the unusual **Straight Range**. As its name implies, it is a line of mountains running some 80 kilometers in a straight line. Despite their rugged appearance these mountains are only about 1,800 meters in elevation. The Teneriffe mountains are somewhat taller, topping out at 2,500 meters. By comparison, the tallest peaks in the Apennine mountains on the east border of Mare Imbrium are close to 6,000 meters above "sea level."

The expanse of Mare Imbrium to the south of Plato contains the largest system of wrinkle ridges on the Moon. This particular system appears to have elements that are broadly concentric with outer perimeter of Mare Imbrium, although many segments do not fit this pattern. The conforming sections, therefore, may mark the remnants of the inner wall of the basin, long since destroyed by volcanic action. If you trace the wrinkle ridges along their full circumference, you'll see that the Teneriffe Mountains and the Straight Range are the lone remains of an enormous basin wall that once was some 800 kilometers in diameter.

## The Alpine Valley

**Vallis Alpes**, the Alpine Valley, in the lunar Alps on the northeast edge of Mare Imbrium, is one of the most attractive formations on the Moon. You first catch sight of it as a dark streak cutting through the hills of the Montes Alpes at First Quarter Moon. As the Sun rises higher it illuminates the edges and floor of Vallis Alpes, revealing much detail. Examination under high power discloses that the valley originates in an ill-defined bulbous area high in the southern portion of the Alps then runs downhill to end at the southern shore of Mare Frigoris. An extremely fine sinuous rille snakes down the middle of the valley, but this feature is only visible through large instruments, and even then only under the best viewing conditions. Much easier to spot are the two transverse faults cutting across the valley. The floor of Vallis Alpes appears smooth, marelike, and (the rille apart) apparently featureless.

Lunar Orbiter photographs show that the side walls are highly irregular, with the eastern wall being much straighter. Only a few craterlets dot the floor, demonstrating its relative youth. The central sinuous rille originates at the head of the valley in a vent crater of a type associated with sinuous rilles elsewhere on the Moon. It then follows a tortuous, meandering course generally down the center of the valley floor. A spectacular oblique view of Vallis Alpes was obtained by Lunar Orbiter 5 looking south toward Mare Imbrium. This photograph shows that the surrounding area was the scene of much volcanic activity: many tiny rilles and volcanic cones can be spotted.

These, unfortunately, are all invisible from Earth. The two faults cutting across the valley clearly extend well into the surrounding highland area. Because this formation lies so far to the north, it was not photographed during the Apollo missions, which didn't orbit north of about 25° lunar latitude.

The Alpine Valley has no exact counterpart on the Moon, and neither the near or the farside has anything closely resembling this valley, which stretches some 150 kilometers and is up to 8 kilometers wide. Its appearance has fostered much speculation about the valley's formation and its geological significance. Theories have ranged from the idea that it represents a chain of volcanic calderas to the scar of a grazing-collision from an asteroid. Another suggestion was that it might represent a valley carved by a glacier! One clue to its origin can be seen when the extreme grazing light of sunrise first brushes across the Alps. At such a time a careful observer is presented with the illusion that several other alpine valleys are lurking deep in the shadows, all cutting across the Alps parallel to the true valley. More than just a trick of the light, this effect underscores the intense fracturing of the lunar crust that occurred during the great Moon-shaking impact that created Mare Imbrium 3.85 billion years ago. The entire area surrounding this sea abounds with radially oriented scars.

The Alpine Valley is actually a graben, a swath of terrain dropped between two roughly parallel faults. Contemporary with its formation, the valley was filled with ejecta. Then flows of lava repaved its surface, giving it the smooth appearance it now has. The rille on this smooth floor is the remains of a lava tube that collapsed when the last of the flows ceased.

## Palus Putredinis

**Palus Putredinis**, the Marsh of Decay, consists of a roughly triangular area bounded by the Apennines to the southeast and the large flooded crater Archimedes at the apex to the northwest. A quick glance at this patch of terrain shows you that the area looks like it has been altered by some interesting geological processes. Montes Apenninus comprise the largest and best preserved lunar mountain chain. It is a portion of the remnant outer wall of the basin created by the asteroid impact responsible for the Imbrium basin. The slope facing Palus Putredinis is quite steep by lunar standards, while the outer slopes, which descend southeastward to Mare Vaporum, are more gradual. These gentle outer slopes consist in part of ejecta from the Imbrium impact. Only a few craters overlie this ejecta, indicating its relative youth.

On the Palus Putredinis side of the mountain range, a series of straight rilles run parallel to the shore. The largest of these is **Rima Bradley**. Just to the north is a set of three rilles, side by side: the **Rimae Fresnel**. These were all formed when the crust under the mare subsided under the weight of the lava in-fill and cracked along pre-existing fractures. Also located in this area is the sinuous rille **Rima Hadley**, the site of the Apollo 15 landing. Unlike the foregoing, Rima Hadley is a volcanic lava tube whose roof has collapsed, exposing layers of mare flooding in its

**The Alpine Valley (below center) is a beautiful sight in small telescopes. Stretching 150 kilometers, the valley is a fracture that filled with lava shortly after it formed. Gerard Therin photo.**

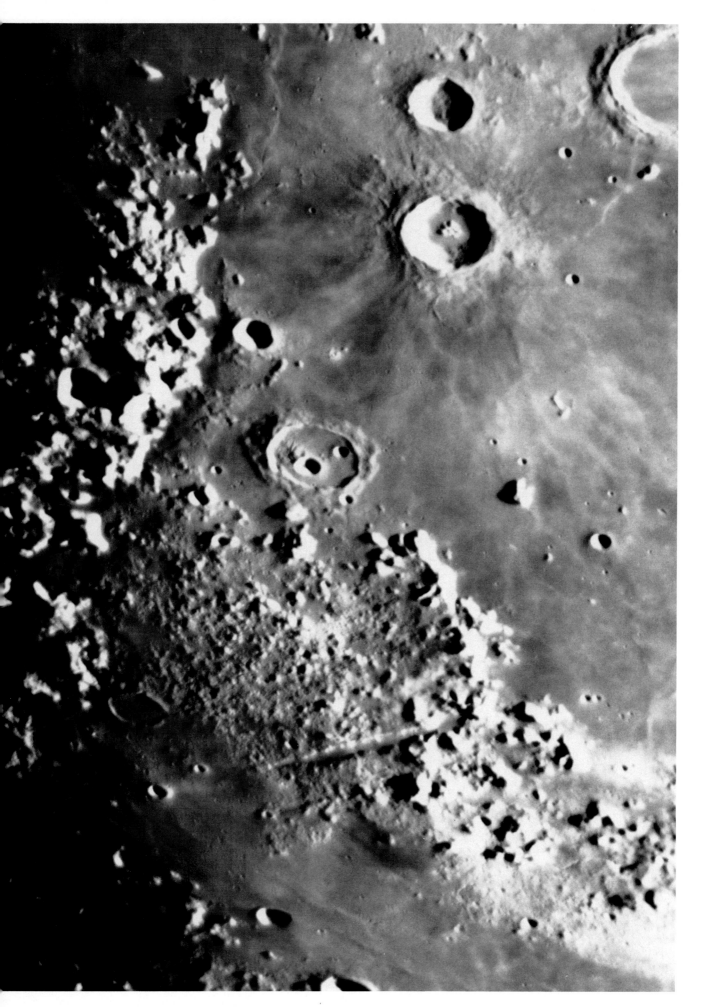

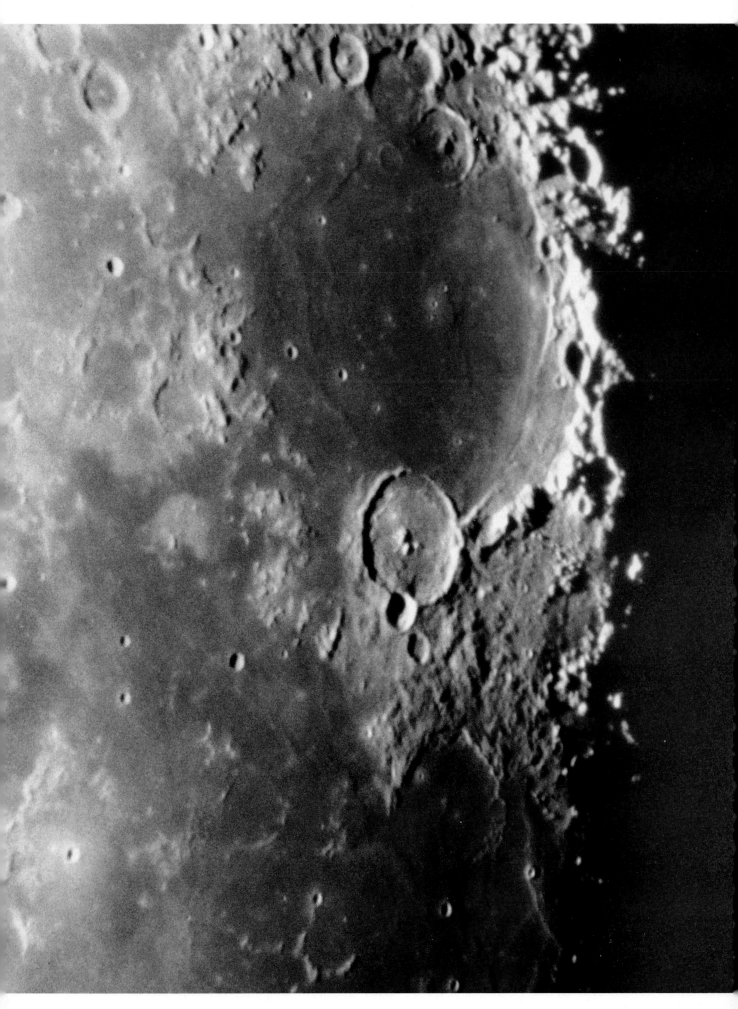

walls. Because of the long shadows cast by the Apennines, which rise 3,500 to 5,500 meters above the plain, Rima Hadley is somewhat difficult to spot at sunrise. By the time it emerges from the shadows, the Sun angle is high enough to wash out some detail. Just before local sunset at Last Quarter is a better time to observe this feature.

The western edge of Palus Putredinis is marked by the fine crater **Archimedes**, noted for its terraces and flooded floor. The floor is so smooth and featureless that a night of steady seeing is needed to spot the two tiny craterlets located near the rim. As the Sun rises higher over Archimedes, its floor will appear to be crossed from east to west by a tracery of rays from Autolycus. The rough terrain of Palus Putredinis is in large part comprised of ejecta from Archimedes. Numerous rilles trend radially away from Archimedes toward the Apennines. Although only two or three of these are easily spotted, as many as eight may be observed when the lighting is favorable.

A careful study of the relationships among the formations here suggests the following sequence of events. An asteroid impact created the Imbrium basin, with the Apennine mountains being a segment of the main rim. A later impact created Archimedes, which at the time must have appeared much as Copernicus does today — with central peaks, a defined ejecta blanket, and a ray system. At some point after the Archimedes impact, large-scale volcanism began filling up the Imbrium basin. Low lying areas, including a good portion of the Archimedes ejecta blanket, were flooded. Lava also infiltrated the floor of Archimedes itself, inundating its central peaks and providing a smooth new surface. As layer upon layer of lava piled up over millions of years, the Imbrium basin sagged under the sheer weight. Cracks appeared at weak points in the crust, both parallel to the Apennine shore and radial to Archimedes, creating the system of rilles in the region. These then are the youngest formations.

## Stadius

High-resolution photography during the Lunar Orbiter and Apollo missions was remarkably effective in resolving lunar mysteries that stumped earlier investigators. Not the least of these unexplainable problems was the structure named **Stadius**. Before the space age, astronomers and planetary geologists were at a loss to explain this odd formation. One popular theory suggested that Stadius was a crater frozen in the act of forming. This theory did not attempt to explain why no other "arrested development" cases could be seen anywhere else on the Moon.

Located in Sinus Aestuum just southwest of the crater Eratosthenes and one crater-diameter due east of Copernicus, Stadius can be viewed to advantage only when the terminator passes over the area and throws fine detail into high relief. This occurs eight days after New Moon and again seven or eight days after Full. Seen at its best, Stadius presents two aspects. Primarily, it is a ghost crater — the remnant of an ancient impact crater, flooded and mostly buried under the smooth lava flows of Sinus Aestuum. To the northeast, one small arc of the crater's rim is preserved; it connects to a group of low hills between Stadius and Eratosthenes. The remainder of Stadius' rim

Circular "seas" are usually old impact basins brimming with now-frozen lava; Mare Humorum is no exception. Around its rim lie craters of all sizes that show signs of being softened and inundated by the mare-filling flows. Lee C. Coombs photo.

consists of nothing more than bits and pieces. However, superimposed on Stadius are numerous craterlets. When you trace the feature's 65-kilometer diameter, you find yourself following a rim alternately composed of low wall segments and craterlets. To be sure, the entire area is peppered by a large number of small craterlets, obviously secondary impacts from nearby Copernicus. Yet when viewing Stadius, you are left with the distinct impression that the number of craterlets superimposed on the rim far exceeds the background density of the others scattered about the area. But measurements made using Lunar Orbiter and Apollo photographs, which resolved many smaller craters not visible from Earth, clearly demonstrate that the craterlets are indeed randomly distributed.

To the northwest of Stadius is the prominent crater chain **Rima Stadius I**. This formation is made of secondary impacts from the Copernicus strike. (Two similar crater chains can be seen to the west near the crater Gay Lussac.) These craterlets almost twinkle if the seeing is poor, and sharpen up marvelously as seeing improves. Were Stadius better preserved, it would form a striking pair with Eratosthenes. Situated at the southwest tip of the Apennine mountains, 60-km **Eratosthenes** is nearly the same diameter as Stadius. Well preserved, its interior walls show complex slumps, and piles of debris and impact melt fill most of its floor around the multiple central peaks. Eratosthenes presents an interesting observing phenomenon. Despite its relative youth, it virtually disappears from view at Full Moon. This is in sharp contrast to other craters of a similar age, which generally have bright rims that stand out clearly under high light — Theophilus, on the shores of Mare Nectaris, is a good example. Also, the appearance of the central peak varies in unusual ways as the lunar month progresses, presenting remarkably different views from night to night.

## Copernicus

On the ninth evening after New Moon, **Copernicus** gradually emerges from the lunar night as the terminator slips across it (photo on page 62). This crater, fresh by lunar standards (only 810 million years old), has all of its features well preserved and is perfectly placed for observation since it lies near the center of the lunar disk. No other crater exemplifies the typical lunar crater better than Copernicus. Long before the morning terminator actually reaches Copernicus the highest peaks on its rim are catching the first rays of sunlight. Over several hours these illuminated points slowly grow wider, then merge to form a ring of light encircling a deep pool of shadow.

Sunrise is the best time to examine the secondary craterlets that radiate outwards from Copernicus (photo on page 61). These tiny craters were gouged by the blocks of rock hurled by the impact that created Copernicus. Raining down on the landscape around the crater, they formed shallow secondary craters best seen in grazing light. Some are arrayed in a herringbone pattern, with the V always pointing back toward Copernicus. At moments of perfect seeing, hundreds of craterlets will suddenly pop into view, providing a breathtaking effect that's well worth waiting for. Closer in to the crater wall you can see the ropy-textured, hummocky ejecta blanket that surrounds Copernicus. The twisted, tortured appearance of this landscape bears evidence of the turbulent outflow of gas and debris that surged from the impact at ground level.

On the tenth night after New Moon, sunlight streaming into the crater bowl will illuminate the three main central peaks. If you time it right, you may catch this scene just as the first rays of sunlight touch the tips of the peaks. Then you can watch the peaks emerge from the long lunar night until, later on, they cast long shadows on the tangled floor of Copernicus. This is a good time to explore the terraces and ravines on the inside of the western wall. These terraces are huge landslides, where miles-long portions of the inner wall collapsed and slumped. At high power a wealth of detail is evident that defies description.

On the eleventh night after New Moon the Sun reveals the floor of Copernicus fully. Note the differences in texture between the northern portion, which is smooth, and the rough and hummocky southern part. As the terminator continues its westward march, you'll see different parts of the complex ejecta blanket and its associated secondary craters. About one crater diameter to the northwest of Copernicus, look for two large crater chains formed by the ejecta.

As Full Moon approaches, Copernicus brightens swiftly and details fade, but the prominent and intricate system of rays that stretch in all directions emerges in all its glory. The force of the impact that created Copernicus and pulverized millions of tons of rock is made apparent by the ray system, which is a titanic splash pattern radiating outwards for hundreds of kilometers. An unusual feature in this ray system are the lovely ''oval rays''. Although most lunar rays run arrow-straight, for some unknown reason a few Copernican rays fell to the ground in two looped patterns, one nested inside the other. Located two diameters southeast of the crater, the interior of the oval rays contains the darkest lava plain in the immediate vicinity of Copernicus. To the west the rays of Copernicus overlap the smaller ray pattern emanating from Kepler.

### Sinus Iridum

Little compares with the beauty of sunrise over **Sinus Iridum**, the Bay of Rainbows, located in the northwest quadrant of the Moon (photo on page 63). Look for this feature 10 days after New Moon and you will immediately be drawn to what (with a little imagination) appears to be the jewel-encrusted handle of a sword projecting out into the unlit velvet black stretches of Mare Imbrium. Each of the sparkling jewels is one of the numerous tips of the **Montes Jura** which encircle Sinus Iridum to the west and north. To the southeast this bay opens to the smooth surface of the mare lava flows. The mouth of the bay is formed by two capes: the northeast point of the Jura Mountains is named **Promontorium Laplace**, while the southwest point is called **Promontorium Heraclides**. Luna 17 landed to the south of Cape Heraclides in November 1970, where it discharged the mobile laboratory Lunokhod 1 onto the surface. Lunokhod drove some 10 kilometers over a ten-month period, taking detailed photographs and carrying out various mechanical and chemical tests.

Sinus Iridum is actually the remains of a large impact crater that formed on the surface of the Imbrium basin before the lava floods came. When they emerged to fill the basin, they engulfed the crater, broke down its seaward wall, and flooded its interior. Stretching across the smooth surface of Sinus Iridum from Cape Laplace to Cape Heraclides is a fairly prominent wrinkle ridge, strikingly visible in the grazing light of sunrise. At least six more wrinkle ridges can be seen within the bay, running parallel to the first. When you see this you can't help but think of a line of ocean waves heading toward a shore. No wonder early observers truly thought the Moon was covered with water (although it soon became obvious that these waves never moved).

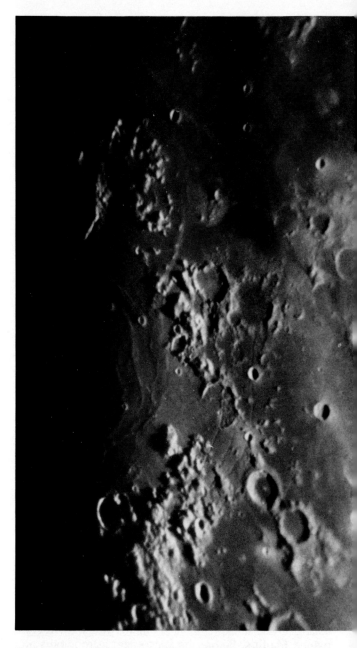

Its floor crisscrossed with rilles, Gassendi offers fine viewing challenges for nights with excellent seeing. How many rilles and craterlets can you spot? Above: The border between Mare Humorum and Mare Nubium shows wrinkle ridges and arcuate rilles. Viewing this region will reveal complicated relationships between faults, craters, ridges, and mountains. Photos by Gérard Therin (right) and Tony Pacey (above).

The mountainous region surrounding Sinus Iridum abounds in features. Just north of Cape Laplace is the much degraded crater **Maupertuis**. Moving counterclockwise around the rim of Sinus Iridum, we encounter in turn the fresher craters **Bianchini, Sharp**, and **Mairan**. Mairan lies due west of Heraclides Promontory, and all four craters are about the same size, 40 kilometers in diameter. Trending northwest from Sharp is a curious small valley which seems to connect Sharp with a smaller crater about one diameter away.

Due south of Mairan on the shore where the mountains meet the sea is the interesting formation **Gruithuisen Gamma**, which resembles an overturned bath tub. This is really a tall, oblong dome about 20 kilometers long. Examine this dome under high power when the seeing is

steady and with an 8-inch or larger telescope you should be able to detect a pit crater at its summit. This tiny (2 km) crater is the vent hole for the dome. Gruithuisen Gamma is one of the few domes with an easily detected pit crater.

Moving back into Sinus Iridum, examine the "shoreline" of this bay. The Jura Mountains plunge precipitously into the sea of lava, but in spots you'll notice some small offshore islands, some spits, and other formations typical of a terrestrial seashore. Why this resemblance? Perhaps it is because the mare lavas had about the consistency of motor oil and flowed much like ordinary water, rather than the typical way the more viscous terrestrial lavas flow.

## MARE HUMORUM

The Sea of Moisture, **Mare Humorum**, might be called the "Sea of Rilles" because of their preponderance around its perimeter (photo on page 66). Humorum is a well-preserved circular mare with not a single large crater on its floor. But a patient observer could count on its surface over 24 small craterlets and dozens of whitish spots which presumably contain even smaller craters. A large and complex wrinkle ridge system arcs north to south on the eastern half of the mare, and on the mare rim at 12 o'clock is the interesting crater **Gassendi**, described below. At least seven rilles meander within Gassendi's walls. Just east of Gassendi, the northeast rim of Mare Humorum has been breached and there its lavas merge with those of Oceanus Procellarum.

At 2 o'clock lies the mostly demolished crater **Agatharchides** and at 3 o'clock is the equally battered **Hippalus**. This locale contains the great system of rilles described below. To the southwest of Hippalus is the isolated massif, **Promontorium Kelvin**, whose steep flanks contain much detail that defies description, especially as the changing angle of sunlight performs a light and shadow show on it. At 5 o'clock we find the smallish but interesting crater **Vitello**, whose floor contains an unusual spiral-like structure, the appearance of which changes dramatically under varying lighting conditions.

Due south of Vitello, deep in the highland area, is the highly elongated crater **Schiller**. Measuring 180 kilometers lengthwise and only 70 kilometers across, Schiller was once thought to be the result of a grazing impact. Lunar Orbiter photography clearly showed that Schiller is actually a compound crater: lava flooding the floors of the two merged craters has erased much of the evidence for the two impacts. Immediately west of Schiller is the vast crater **Schickard**, whose flooded floor is crossed by light markings. And directly behind Schickard is one of the strangest craters on the Moon, **Wargentin**. These two formations are covered in detail on page 72.

Let's return our attention north to Mare Humorum. Just west of Vitello is the mostly ruined crater **Doppelmayer**, 65 km in diameter. It is interesting to trace the nearly destroyed rim of Doppelmayer around its full circumference, noting how the encroaching lavas from Mare Humorum have effected various amounts of destruction. Just northeast of Doppelmayer is **Puiseux**, an even more dilapidated crater that almost became a ghost. The western side of Mare Humorum presents another concentration of rilles complementary to those near Hippalus. A careful observer with a moderate aperture telescope might be able to detect 20 rilles or more in this region. The most important are highlighted on page 72, which also covers the noteworthy crater **Mersenius**.

### Gassendi, Letronne, and Flamsteed

**Gassendi** is one of the most significant and interesting

craters on the Moon. Measuring 110 kilometers in diameter, it was the runner-up landing site for Apollo 17, losing to the Taurus-Littrow valley east of Mare Serenitatis. Gassendi dominates the northern rim of Mare Humorum and is a spectacular object. The much younger crater **Gassendi A** is superimposed on its rim, giving the appearance of a diamond ring at sunrise, which occurs about 11 days after New Moon. Sunrise is a good time to begin observations, for the floor of this crater contains an unusual wealth of detail that is best revealed under low illumination. You become immediately aware of the rough, jumbled texture of the floor. In the center is a group of four peaks. But the highlight is a magnificent system of rilles which crisscross the eastern half of the crater's floor.

On the northwest portion of the rim are two curious notches, easily spotted. To the south, where Gassendi protrudes into Mare Humorum, the crater wall has been broken and leveled. Lavas from the adjacent mare flowed through this gap and flooded a small crescent of Gassendi's floor, which appears much smoother to the eye. Gassendi is an area where many transient glows have been reported by experienced observers, perhaps due to venting of gas through the many rilles found there.

Directly north of Gassendi is the semi-ruined crater **Letronne**, 120 km in diameter. The northern portion of this formation has been completely destroyed by the encroaching lavas of Oceanus Procellarum, creating a bay-like feature. In fact, after Sinus Iridum Letronne is one of the finest examples of a lunar bay. In the center of this ruin you can just make out the tiny protruding tips of three central peaks that are almost, but not quite, buried under the flood of lava. In contrast to Gassendi, the floor of Letronne is smooth and almost featureless. If observed at local sunrise or sunset, you can detect a system of wrinkle ridges overlying the area where the now-missing crater wall once stood.

Moving further north and slightly west, we come across the small crater **Flamsteed**, which sits just within the rim of an ancient and all-but-obliterated ghost crater designated **Flamsteed P**. Within this 112-km ring is the landing site of the unmanned Surveyor 1 surface probe. All that remains of Flamsteed P are bits and pieces of its old crater rim — everything else was long ago ruined by lava flows. This ghost ring is best seen just as the Sun rises over it. A few days later it disappears from view, but then it begins to brighten again and becomes distinct at Full Moon.

Within a small area we have explored three craters of similar size but of totally dissimilar appearance due to the action of lunar lava flows. Gassendi retains most of its original features. Letronne has been destroyed on its seaward side, yet the remainder shows its similarity to Gassendi. Flamsteed P is so degraded as to be barely recognizable as a crater, yet it once was as magnificent as Gassendi.

### Rimae Hippalus

The eastern rim of Mare Humorum is the site of what many consider the most spectacular set of rilles on the Moon. Unusually well preserved and running a great distance, the **Hippalus rilles** are easy to spot and observe when the Sun is rising over them about 10 days after New

**A lunar oddity, Wargentin is a crater that was filled to the brim — and perhaps overflowing — by lava. Down its center runs a split wrinkle ridge that's a challenge to spot unless lighting conditions are just right. Lick Observatory photo.**

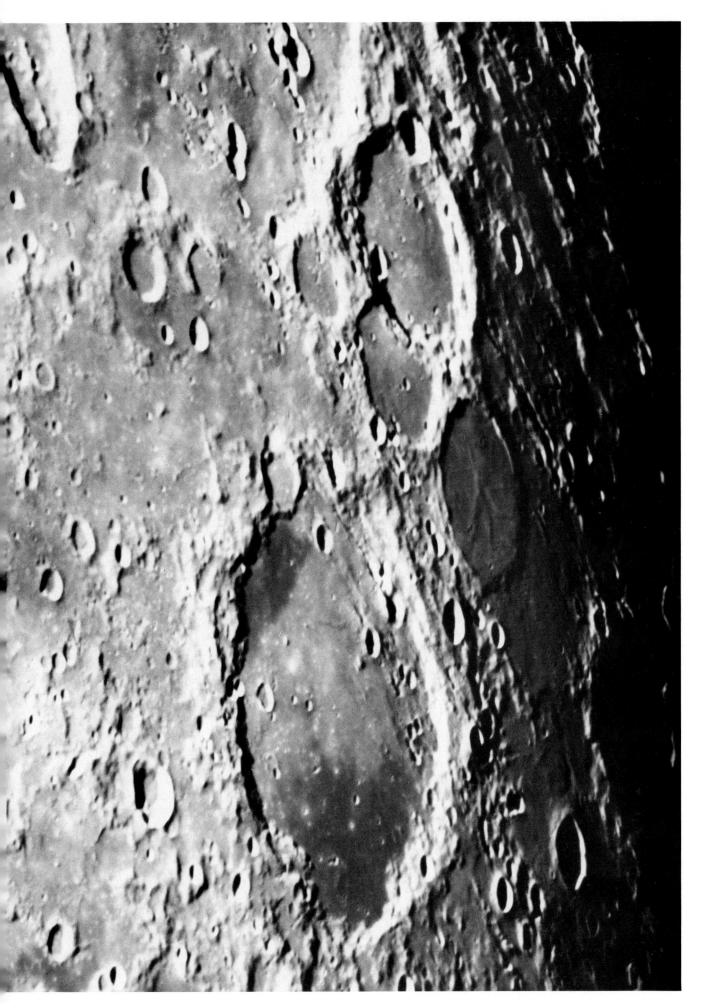

Moon. Rimae Hippalus are the foremost example of parallel grabens, and the longest of these arcuate rilles extend over a quarter of the circumference of Mare Humorum. The most clearly visible portion of this arc is located in the area between the half-destroyed crater **Hippalus** and the prominent crater pair **Campanus** and **Mercator**, which are located on the southwestern border of Mare Nubium.

One rille bisects the crater Hippalus, a second runs parallel to the first and lies tangent to the east wall of Hippalus, and the third major rille lies further to the east by an equal distance. A shorter and somewhat thinner rille can be seen between the first two. The three major rilles run north past Hippalus for another 60 kilometers, ending near an ancient structure designated **Agatharchides P**. In the other direction, these rilles can be traced south far into the hilly region east of the crater Vitello.

A number of other rilles populate the area. One can be found on the smooth floor of Agatharchides P. Another rille runs directly between Campanus and Mercator; this one parallels the Hippalus rilles. This rille ends to the south in a small patch of mare called **Palus Epidemiarum**. A complicated system of short rilles is located around the crater **Ramsden** in this mare area. Immediately to the east of Ramsden, two of these rilles overlap in an X-shaped pattern. In this area, it is worthwhile to examine the crater **Capuanus**, located just east of Ramsden on the southern portion of Palus Epidemiarum. At sunrise, the floor of Capuanus will be seen to contain as many as six domes. This is one of the few examples of domes to be found on a crater's floor.

## Schickard and Wargentin

The mountainous region along the southwestern limb of the Moon contains two fascinating objects. One of the largest craters on the Moon, **Schickard** is our guidepost to what is probably the most unusual crater of all: **Wargentin**. Nestled in a jumbled highland crater field, Wargentin could be difficult to locate if it weren't adjacent to the immense Schickard, fully 230 kilometers in diameter. Schickard's floor has been flooded with lava and its smooth appearance sets it off from the surrounding area as it comes into view just before Full Moon.

The interesting feature of Schickard is the strange pattern of light and dark material on its floor, which intensifies as the Sun rises higher. Schickard appears to be highly elongated due to the effects of perspective because it lies so near the lunar limb. In reality, it is nearly circular. The two lobes of Schickard are quite dark, while a V-shaped wedge of lighter material dominates the center of the crater floor. Perhaps if Schickard lay further from the limb we could determine the cause of the coloration. But it will probably take a lunar orbiter survey spacecraft to settle the matter.

Once you have learned how to find Schickard, spotting Wargentin immediately to the south will be easy. Although it is 84 kilometers in diameter, Wargentin appears much smaller because of its proximity to the limb. However, the steep angle at which we see the highly foreshortened Wargentin shows its main features off to great advantage. This is by far the largest of an extremely rare class of highlands impact crater which has been filled with lava. Flooded craters are nothing special in the mare areas, but in the case of Wargentin, you see a highlands crater that lava has filled right up to the rim, giving it the appearance of a mesa.

The best time to observe Wargentin is about 12 or 13 days after New Moon, when the Sun is rising on this for-

mation. (You can't observe Wargentin at local sunset, because this occurs only three days before New Moon, when the Moon is a sliver low in the morning sky.) As the terminator crawls across its flooded floor, you'll see a remarkable wrinkle ridge shaped like a Y. Wrinkle ridges are common on the lunar seas, but the presence of one inside a crater is unprecedented. This gave astronomers the clue pointing to a common origin for the floor of Wargentin and the lunar seas — namely the flow of lava extruded from within the Moon's mantle. High-power observation as local sunrise continues will reveal more wrinkle ridges and a host of other features that belie the smooth appearance Wargentin possesses at low powers.

Moving around the edge of this odd crater, note that a piece of the rim still remains on the eastern side. No trace of wall can be found elsewhere; presumably lava flowed over the rim, completely obliterating it. A day or two after sunrise Wargentin vanishes into the jumble of craters in this area, although its location is clearly pinpointed by the obvious dark and light pattern of Schickard.

## Mersenius

The crater **Mersenius** is notable for its convex floor and for its association with a fine set of rilles. Located just west of Mare Humorum, it is a well-formed crater 82 kilometers in diameter, just slightly smaller than Copernicus. When the terminator passes across the flooded floor of Mersenius on the twelfth day after New Moon the convexity of its floor is readily apparent. No other crater on the Moon displays the effect so obviously. The convexity results from large-scale doming of the floor due to the upwelling of magma beneath it. The proximity of Mersenius to Mare Humorum lends credence to this hypothesis, because flooding from beneath is commonly seen in craters that rim the lunar maria.

East of Mersenius on the floor of Mare Humorum is a set of three arcuate rilles which complement the Hippalus rilles on the other side of the sea. Both undoubtedly have a common origin. The largest of the Mersenius rilles is the one closest to the crater. It can be traced north to a point where it moves off the mare floor and into the surrounding highland terrain. There it extends for at least another 100 kilometers. The middle rille of the three is the shortest and least distinct. The third rille, traced south of Mersenius, converts to a fault. This effect is also seen near the crater Cauchy in Mare Tranquillitatis. The fault, called **Rupes Liebig**, is seen as a bright thin line on the twelfth day after New Moon. This shows that its wall faces east, thereby catching the first rays of the rising Sun.

Exactly one crater-diameter south of Mersenius is the much eroded and flooded crater **de Gasparis**, which is the focal point for a collection of nine short rilles. Particularly interesting is the fact that de Gasparis and an identically sized crater butted up against it to the south, **de Gasparis A**, are both crisscrossed by rilles forming X-shaped patterns on their floors.

## OCEANUS PROCELLARUM

The western hemisphere of the Moon is dominated by one vast mare, **Oceanus Procellarum** — the Ocean of Storms. Opinion is still divided as to whether Oceanus Procellarum is the result of a single huge basin-forming impact, or if it represents the coalesced remains of several smaller basins. (To settle the question may take detailed geological traverses across the feature.) In the north it merges with Sinus Roris, a bay that connects Mare Frigoris

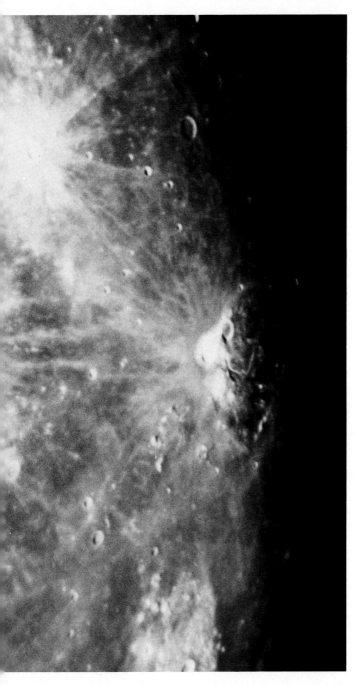

Lying in northwest Oceanus Procellarum, the region around Aristarchus is a delight to explore by telescope. Notice how the rays from Aristarchus mingle with those from Kepler at upper left and from Copernicus at left. K. Ken Owen photo.

to Oceanus Procellarum. In this region, due west of Sinus Iridum, we come across the unusual volcanic construct, **Rümker**.

A solitary complex of lunar domes some 55 kilometers across, Rümker is best observed at sunrise, twelve days after New Moon. By the time of Full Moon, virtually all detail is lost because of the low relief of this structure. Whereas all the known domes on the Moon are solitary or in small groups of individuals, here we find the only example of domes clustered on top of other domes, giving Rümker a rumpled surface. Careful examination will reveal numerous vent craterlets.

Moving south across featureless expanses of frozen lava, we first encounter the **Aristarchus** uplift, a lozenge-shaped geological area covered in detail on page 74. To

the east of the uplift are the **Montes Harbinger**, which each month presage the appearance of Aristarchus and Schröter's Valley. Near these mountains is the remains of a flooded crater named **Prinz**. A series of sinuous rilles originate just north of Prinz; in each case a largish vent crater at the head of the rille can be seen. North of Prinz is the crater **Krieger**, its wall broken on the southern side by a smaller crater. Several wrinkle ridges run through Krieger. Based on evidence of rilles and other surface features photographed during Apollo missions, this crater is believed to be one of the largest volcanic craters on the Moon.

Due west of the Aristarchus uplift, very close to the lunar limb, are the magnificent remains of three enormous craters: **Struve**, **Russell**, and **Eddington**. Long ago flooded by the lavas of Oceanus Procellarum, all that remains of these three are incomplete arcs of walls. Struve and Eddington have seemingly merged into a single formation spanning 300 kilometers, but this is essentially an effect of perspective. What makes this area so beautiful is the extreme angle at which we look sideways into these ruined craters.

Directly south of Eddington are the smaller craters **Krafft** and **Cardanus**, notable for the wide rille, **Rima Krafft**, which runs north-south and appears to connect them. Due east of Cardanus the flooded crater **Marius** is located in a mostly featureless area. Marius forms the northern point of a triangle with the slightly smaller craters **Kepler** (to the southeast) and **Reiner** (to the southwest). Marius is an undistinguished crater some 40 kilometers across which has been flooded with lava, giving it a featureless floor, save for the tiny craterlet **Marius G**. A mere 3 kilometers in diameter, this craterlet is a good test of seeing conditions. But lying immediately to the west of Marius is an extensive field strewn with scores of volcanic domes, which can be seen about 11 days after New Moon, when the terminator is sliding across the dome field.

At this time a scan through the Marius dome field reveals a multitude of shapes — some round, others elongated. A small number have summit pit craters, through which volcanic material was vented. However, at least an 8-inch refractor or a 12-inch reflector will be required to detect the vents. The volcanic nature of this field will be quite obvious, especially to anyone who has visited terrestrial volcanic dome fields. This area is so geologically diverse and interesting that it was considered as the target for Apollo 18 or 19, two missions unfortunately cut from the program. To the southwest of Marius is the lonely crater Reiner, signpost for the mysterious swirl of light material known as **Reiner Gamma**, discussed on page 74.

One of the darkest areas on the Moon can be found on the western limb just outside the edge of Oceanus Procellarum any time after Full Moon. In fact, the flooded basin named **Grimaldi** is so dark that it can be easily spotted on the earthlit side of a 4- or 5-day-old Moon as a ghostly dark patch. Located close to the lunar equator, the flooded dark area of this ancient formation is some 220 kilometers in diameter. Exterior to this is an outer wall, traceable with some difficulty, which nearly doubles the size of Grimaldi. There is no ready explanation for why the lavas that resurfaced the floor of Grimaldi's inner basin should be so dark relative to the nearby lava fields of Oceanus Procellarum.

Immediately to the northwest of Grimaldi is the large crater **Riccioli**, measuring 150 kilometers from rim to rim.

Its walls are much better preserved and Riccioli also has a small dark patch of lava on the northern end of its floor. The general area surrounding Grimaldi and Riccioli is rich with rilles, which can be seen in grazing sunlight a day or two before Full Moon. Due north of Grimaldi is the crater **Hevelius**, 120 kilometers in diameter. Hevelius forms an attractive pair with the much younger and well-preserved **Cavalerius**, only half its size. Hevelius contains an unusual group of intersecting rilles on its floor, but they are small and require a large telescope and good seeing conditions for their detection. The area immediately to the northeast of Hevelius and Cavalerius has been named **Planitia Descensus** in honor of the first soft lunar landing, which was made by the Russian probe Luna 9 in 1966. All these craters are greatly foreshortened into elliptical shapes thanks to perspective effects.

The southern border of Oceanus Procellarum contains two craters, **Billy** and **Hansteen**, which are useful points of orientation. Billy (the southeastern member of the pair) is a flooded crater with a dark floor, which makes it stand out dramatically against the light highland background. Hansteen, identical in size to Billy at 45 kilometers, has numerous hills on its floor and a wide rille located just outside its western rim. In between the two craters is the extremely bright massif, **Hansteen Alpha**. Directly west of Billy and Hansteen is the huge highland rille, **Rima Sirsalis**, which is discussed on page 77.

Returning northwards to the lunar equatorial region, Oceanus Procellarum contains the important crater **Kepler**, center of a bright system of rays which fan out and intersect with those of Copernicus to the east. Kepler is unusually prominent for its size (only 32 kilometers) and has considerable detail on its floor worthy of study. To its south is **Encke**, almost identically large but somewhat older and consequently more degraded in appearance. Yet Encke also possesses much fine detail on its floor.

Midway between Kepler and Copernicus lies an area containing some of the best known groups of lunar domes. The first is located immediately north of the bowl-shaped crater **Hortensius**, the southernmost of the only two sharp rimmed craters in this region. There are six small, perfectly formed domes arranged in pairs. Five of the six have tiny, but well-defined summit craters. To the northwest is **Milichius**, the other bowl-shaped crater. Just to its west is the dome **Milichius Pi**, which also has a pit crater at its summit.

Moving due north from Milichius across a smooth patch of mare, we soon encounter some isolated mountain massifs within a hilly area. These are the western outliers of **Montes Carpatus**. The largest massifs, which are elongated in the southwest-northeast direction, are named **Tobias Mayer Alpha** and **Zeta**. Interspersed among these massifs is another grouping of a half dozen or so domes. Unlike their counterparts near Hortensius, these are broader and lower and are therefore somewhat more difficult to spot, except immediately after local sunrise or just before local sunset. Summit craters can be detected on two of the domes when seeing conditions are excellent. In one case three of these flat domes seem to have nearly merged together.

### Aristarchus and Schröter's Valley

The most fascinating and geologically diverse area on the Moon stands near the western edge of Mare Imbrium. This area is highlighted by the presence of the crater **Aristarchus**, one of the youngest craters on the Moon. Because the violent impact which created Aristarchus occurred so recently, the pulverized rock is still extremely bright. In fact, Aristarchus is the brightest object on the Moon. As a consequence, it is one of the few features that can be readily distinguished when lit only by earthshine. Look for it a few days after New Moon. Almost opposite to the thin sunlit lunar crescent, near the western edge of the disk, you will see Aristarchus and the plume of crushed rock that was tossed from the crater during impact, looking much like a ghostly comet with a stubby tail transiting in front of the Moon. As the lunar crescent grows and earthshine weakens, Aristarchus vanishes and we are forced to wait until about three days before Full to resume observations of it.

Sunrise on Aristarchus reveals a crater with exquisitely defined features: a knife-edge rim, many narrow terraces stepping down to the floor of the crater, and a tiny central peak. The intense brightness of this crater makes the use of a polarizing filter handy. Many observers overlook the central peak because it is overwhelmed by the glare. Also worth looking for are the series of dark bands, caused by landslides, that run vertically down the inner crater wall.

As the terminator moves across the Moon, it reveals a trapezoidal area of complex geology immediately to the north of Aristarchus. This hilly, jumbled terrain, sometimes called the Aristarchus uplift, is set in the midst of extensive smooth lava plains. Larger telescopes and steady atmosphere reveal great geological diversity in the form of numerous crater pits, winding rilles, and domes throughout the immediate area.

But the key feature in this area can be seen in the smallest telescope. This is the famed **Schröter's Valley**, a sinuous rille that originates in an elongated crater aptly called the Cobra's Head, then snakes its way 175 kilometers across the uplift. Schröter's Valley is an enormous collapsed lava tube and is the only sinuous rille on the Moon you can observe easily. All the other sinuous rilles (including the well-known Hadley Rille in Mare Imbrium which was visited by the Apollo 15 astronauts) are difficult objects by comparison.

Lying next to Aristarchus is the crater **Herodotus**, about the same size but otherwise as different as possible. Where Aristarchus is bright and fresh, Herodotus is dark and degraded. Its walls have been much eroded and dark lava has invaded the crater and flooded its floor, obliterating its central peak and the terracing of the crater walls.

### Reiner Gamma

**Reiner Gamma** is not a crater, a dome, a mountain, or a rille. It is instead a swirl of bright material lying in the western reaches of Oceanus Procellarum near the lunar equator. It appears to be a splash of something with absolutely no vertical relief. Reiner Gamma is probably made of the same kind of material as the bright rays systems because it brightens in exactly the same way as the Sun rises higher. However, lunar rays usually run straight and far — moreover, you can trace rays back to the source crater. But in the case of Reiner Gamma, no such identification is possible.

This formation remained a total mystery until photo-

**So bright it "glows" even by earthlight, the white interior of Aristarchus nonetheless shows darker shadows which are scars of landslides on its interior walls. Schröter's Valley is probably a collapsed lava tube. Dragesco photo.**

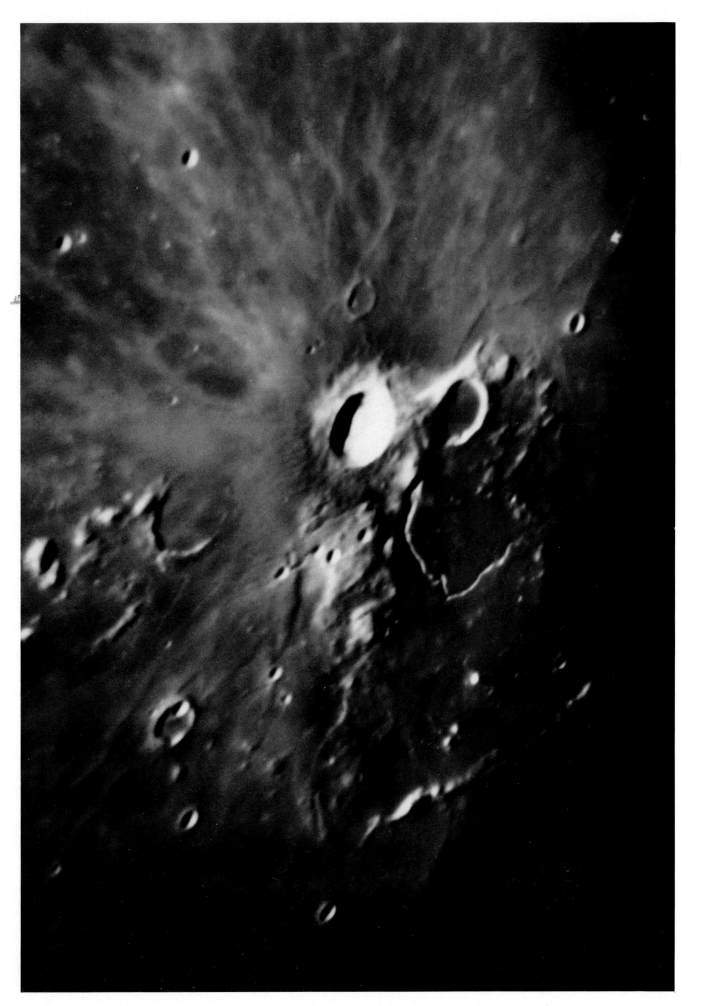

graphs from Lunar Orbiter revealed a second area of swirl deposits in Mare Marginis on the Moon's eastern limb. The source of this much larger swirl perplexed lunar geologists until they noticed that the Mare Marginis swirls were located almost exactly on the opposite side of the lunar globe from Mare Orientale. The Mare Marginis swirl formation was likely created when some of the ejecta from the huge impact that formed Mare Orientale travelled completely around the Moon on low orbital paths. Arriving at the lunar antipodes from all directions at once, this ejecta collided and some of it was deposited on the ground.

Is this the way Reiner Gamma was formed? We can't really be sure. For one thing, there is no identifiable source for the Reiner Gamma swirl. But at least we have an idea of how a formation with the characteristics displayed by Reiner Gamma can be created.

### Rima Sirsalis

The southwest quadrant of the Moon contains the largest rille on the lunar surface. **Rima Sirsalis**, measuring 420 kilometers from tip to tip, is the longest by quite a margin. Located northwest of Mare Humorum, the focal point for the greatest collection of rilles on the Moon, the Sirsalis rille is further distinguished by being one of the few major rilles running deep into highland terrain. Most other rilles are located near the borders of lunar seas.

Rima Sirsalis is a fine example of the graben type of rille. It is best observed about 12 to 14 days after New Moon, when the Sun angle is low and grazing light throws it into dramatic relief. It then appears as a slender dark line, clearly visible in the smallest telescope. It can also be observed on the waning crescent Moon some 4 to 5 days before New Moon.

The Sirsalis rille originates at the edge of Oceanus Procellarum near the splendid bay **Sirsalis E**. From that point it trends southwest past the distinctive crater pair **Sirsalis** and **Sirsalis A**. Sirsalis is a fresh crater measuring 44 kilometers in diameter which overlies an older, but identically sized crater. This striking pair is an excellent guide to the rille. Just south of Sirsalis the small crater **Sirsalis J** stands on top of Rima Sirsalis, interrupting its path for a short distance.

Continuing to the southwest, Rima Sirsalis next passes to the east of the crater **Crüger**. This is one of a number of craters in this area, including Zupus and Billy, having very dark floors. Although Crüger is almost identical in size to Sirsalis, the contrast between its dark floor and the bright surrounding highlands accentuates its size, giving it the illusion of being much larger than it really is.

The Sirsalis rille then passes across the walls and floor of a crater designated **de Vico A**. It is interesting to observe in larger instruments how the younger rille overlies this older crater. Finally, the cleft passes into an ancient and degraded unnamed crater adjacent to the eastern wall of the highly disintegrated walled plain **Darwin**. Here Rima Sirsalis is remarkably crisscrossed by perhaps a half dozen older rilles, the longest of which extends all the way into the crater Darwin. This overlapping of rilles is another unusual aspect of Rima Sirsalis.

**Barely visible from earth, Mare Orientale is a bull's-eye-shaped impact basin straddling the Moon's western limb. The dark markings are lava flows that have partially flooded the lowlands. Drawing by David L. Coleman.**

If the Sirsalis rille were located near the center of the lunar disk it would be ranked as one of the most spectacular sights on the Moon. As it is, even much foreshortened by its location near the western limb, it is still clearly of great length, although its true width is not at all obvious. Nonetheless, its many remarkable features will reward your hours of careful observation with breathtaking views.

## MARE ORIENTALE

Space photography has revealed **Mare Orientale** — the Eastern Sea — as the most spectacular formation on the Moon. Unfortunately, this feature sits astride the Moon's western limb (the sea's name dates from the old convention regarding lunar east and west). Because of this we can never view it in its entirety from Earth. Enough of Orientale is revealed at certain times, however, to provide us with some provocative glimpses. The youngest and best preserved of all the major lunar basin-forming impacts — it is 3.8 billion years old — Mare Orientale looks precisely like a bullseye when seen from directly above, a vantage point only experienced by the Lunar Orbiters and the Galileo spacecraft on its way to Jupiter. At best, we can see slightly more than half the bullseye from Earth and then only at an oblique angle. This occurs when favorable librations in both latitude and longitude combine to swing Mare Orientale more toward the center of the lunar disk.

To see a sunrise over Mare Orientale, which throws its rugged basin walls into high relief, you must observe it just before Full Moon. Every month, some portion of the Orientale basin walls will be silhouetted on the lunar limb. Some isolated peaks project well above the circular outline of the Moon, attesting to the formation's ruggedness. However, under these conditions of lighting, it is difficult to appreciate the true form of Mare Orientale. On infrequent occasions a combination of maximum favorable librations in latitude and longitude occur simultaneously at the time of sunrise over Mare Orientale. These bring nearly 60 percent of the formation onto the nearside and you can begin to detect the bullseye form. The last of these most favorable apparitions occurred during a six-month period centered around November 1985; the next set of optimum conditions occur in November 2003. Still, fine views can be had at other times. The best way to judge whether librations are favorable is to glance at Mare Crisium, on the opposite side of the Moon, just prior to Full Moon. If Crisium is swung way over near the eastern limb of the Moon, conditions for observing Mare Orientale will be favorable.

It is also instructive to observe Mare Orientale at local midmorning, beginning a few days after Full Moon. At this time the mare areas darken and delineate the shape of this formation, although you lose all sense of relief. When librations are favorable, the dark patch located at the center of Mare Orientale can be seen right on the edge of the Moon. To the northeast, two narrow strips of dark mare material mark the inner and outer rings of the bullseye. The inner strip of mare is called **Lacus Veris**, while the outer one is named **Lacus Autumnae**. The ring of mountains in between Lacus Veris and Lacus Autumnae is **Montes Rook**; the outermost basin wall is **Montes Cordillera**. You should be able to detect the relief created by both these mountainous regions as they pass across the limb of the Moon and onto its invisible farside.

# Bibliography

**Alter**, Dinsmore. *Lunar Atlas.* 343 pp., paper. Dover Publications, New York, 1968.

**Alter**, Dinsmore. *Pictorial Guide to the Moon.* 216 pp., paper. Thomas Crowell, New York, 1973.

**Cadogan**, Peter. *The Moon; our sister planet.* 391 pp., hardcover. Cambridge University Press, New York, 1981.

**Cherrington**, Ernest H., Jr. *Exploring the Moon through Binoculars and Small Telescopes.* 229 pp., paper. Dover Publications, Inc., New York, 1984.

**Cortwright**, Edgar M., ed. *Apollo Expeditions to the Moon.* [NASA SP-350]. 313 pp., hardcover. U.S. Government Printing Office, Washington, 1975.

**French**, Bevan M. *The Moon Book; exploring the mysteries of the lunar world.* 287 pp., paper. Penguin Books, New York, 1977.

**Hartmann**, William K. *Moons and Planets.* Second ed., 509 pp., hardcover. Wadsworth Publishing Co., Belmont, California, 1983.

**Hill**, Harold. *A Portfolio of Lunar Drawings.* 240 pp., hardcover. Cambridge University Press, New York, 1991.

**Kopal**, Zdenek. *A New Photographic Atlas of the Moon.* 311 pp., hardcover. Taplinger Publishing Co., New York, 1971.

**Mazursky**, Harold, G.W. Colton, and Farouk El-Baz, eds. *Apollo Over the Moon; a view from orbit.* [NASA SP-362.] 255 pp., hardcover. U.S. Government Printing Office, Washington, 1978.

**Moore**, Patrick. *The Moon.* 96 pp., hardcover. Rand McNally & Co., New York, 1981.

**Moore**, Patrick. *New Guide to the Moon.* 320 pp., hardcover. W.W. Norton and Co., New York, 1976.

**Rukl**, Antonin. *Hamlyn Atlas of the Moon.* 224 pp., hardcover. Paul Hamlyn Publishing Co., London, England, 1991.

**Schultz**, P. *Moon Morphology.* 626 pp., hardcover. The University of Texas Press, Austin, 1976.

**Taylor**, Stuart Ross. *Lunar Science, a post-Apollo view; scientific results and insights from the lunar samples.* 372 pp., hardcover. Pergamon Press, New York, 1975.

**Wilhelms**, Don E. *The Geologic History of the Moon.* [U.S. Geological Survey Professional Paper 1348.] 302 pp., paper. U.S. Government Printing Office, Washington, 1987.

# INDEX

**Photographs are in boldface.**